MORTAL AND
IMMORTAL DNA

MORTAL AND IMMORTAL DNA

Science and the Lure of Myth

GERALD WEISSMANN

BELLEVUE LITERARY PRESS
New York

First published in the United States in 2009 by
Bellevue Literary Press, New York

FOR INFORMATION ADDRESS:
Bellevue Literary Press
NYU School of Medicine
550 First Avenue
OBV 640
New York, NY 10016

Copyright © 2009 by Gerald Weissmann

This book was published with the generous support of
Bellevue Literary Press's founding donor the Arnold Simon Family Trust,
the Bernard & Irene Schwartz Foundation
and the Lucius N. Littauer Foundation.

Library of Congress Cataloging-in-Publication Data

Book design and type formatting by Bernard Schleifer
Manufactured in the United States of America
FIRST EDITION
1 3 5 7 9 8 6 4 2
ISBN 978-1-934137-16-1 tp

Cover image: *Historia naturalis Ranarum nostratium,* v1 p 1: Roesel von Rosenhof (Haarlem, Neth'ds: C.H. Boh; 1764-68)
Used with permission MBL/WHOI Library, Woods Hole MA

August Johann Roesel von Rosenhof (1705-59), descendant of a noble Austrian family, was a miniaturist painter, naturalist and entomologist. His specialties were insects, amphibians and reptiles; these became subjects of some of the most
important works of 18th century natural history.
    Roesel's respect for the wonders of nature is expressed in this book's cover, the frontispiece of Roesel's magnum opus.
A quote by Pliny the Elder (23AD-79AD) takes center stage and translates roughly as "the majesty of Nature lies in its
smallest details." Roesel adds an 18th century Enlightenment fillip. Science, the figure on the right (with Pliny's image
hanging overhead), holds measuring tools and a triangle with a gravitational ball—a nod to Newton. The muse on the
left embodies ideals of learning (the book), its grounding in the real world (her right foot on a globe), peace (the palm
frond), the vision of enlightenment (the sun) and, perhaps, the "naked truth." The putti in the foreground are exploring,
collecting and organizing their marine specimens into a miniature cabinet of curiosities. Presiding over the scene, atop
Pliny's plinth, is Astarte, the Phoenician goddess of fertility, beauty and love.
                    (text by Ann Weissmann, exhibitions curator, MBL/WHOI Library, Woods Hole MA)

For Ann, the first love . . .
and the first reader.

Contents

Science is the great antidote to the poison of enthusiasm and superstition.

—ADAM SMITH, *The Wealth of Nations*

MYTHOLOGY, n. The body of a primitive people's beliefs concerning its origin, early history, heroes, deities and so forth, as distinguished from the true accounts which it invents later.

—AMBROSE BIERCE, *The Devil's Dictionary*

What is serious is not that so many people believe in astrology, but that they take seriously the ideas of those who believe in astrology.

—JEAN ROSTAND, *Nouvelles pensées d'un biologiste*

Prefatory Note

I N HIS HISTORIC *The Two Cultures and the Scientific Revolution* (1959), C. P. Snow described a wide gulf between the two cultures of science and the humanities. He defined two very different worlds: one inhabited by folks who are able to sort out Plato's Myth of the Cave and another by those who can recite Newton's second law of thermodynamics.

> *There have been plenty of days when I have spent the working hours with scientists and then gone off at night with some literary colleagues. . . . If I were to risk a piece of shorthand, I should say that scientists naturally had the future in their bones.*[1]

Writing at the dawn of the atomic age, Snow was concerned that not having "the future in their bones" might lead scientifically illiterate folk to obliterate the world by nuclear mischief. But nowadays, the gulf between the arts and the sciences has been filled by the silt of popular culture, and the mischief people fear is that of experimental biology. Many in the most *avant* of the literary *garde* have turned from math to myth, while many of the folk doing the newest of science haven't looked at a line of verse since college. I share with W. H. Auden the recollection that "my father was both a physician and a scholar so I never got the idea that art and science were opposing cultures—both were entertained equally in my home."[2] The essays in this book are addressed to those interested in keeping the entertainment going.

1. Mortal and Immortal DNA: Craig Venter and the Lure of "Lamia"

From Albertus Seba's *Cabinet of Natural Curiosities* (1734–1765)

"Lamia." From Edward Topsell's *History of Fourfooted Beasts* (1607)

First Individual Diploid Human Genome Published by Researchers at J. Craig Venter Institute. Sequence Reveals that Human to Human Variation is Substantially Greater than Earlier Estimates.

——Press Release, September 15, 2007[1]

This theory suggests that only the differentiating cells will inherit the newly "photocopied" DNA strands, which contain errors. The stem cells retain the unmodified original DNA strands, which consequently remain "immortal" after repeated cell divisions. ——G. Cossu and S. Tajbakhsh, *Cell*, 2007[2]

The capacity to blunder slightly is the real marvel of DNA. Without this special attribute, we would still be anaerobic bacteria and there would be no music.

——Lewis Thomas, 1979[3]

DNA STRIKES BACK

WHEN NEWS ARRIVED THAT THE 2007 NOBEL PRIZE IN PHYSIOLOGY or Medicine had been awarded to Mario Capecchi, Martin Evans, and Oliver Smithies for targeting genes in mice, it crowned the comeback of DNA.[4] For two decades now, fans of RNA have been lording it over those who study its deoxygenated sibling. Ever since Wally Gilbert announced the "RNA world"[5] and Altman and Czech discovered its enzymatic properties,[6] RNA has been flying high. First, RNAi became the molecule of the year,[7] in

2006 RNAi picked up its glittering prize in Stockholm[8] and pretty soon RNAi was roiling the stock market.[9] That hadn't happened to DNA for a while. Like the Korean war, Crick and Watson's double helix goes back to 1953; Meselson and Stahl worked out the chemistry of its replication in 1958; Roberts and Sharp defined the ins and outs of the DNA lexicon back in 1977.[10] Even the Capecchi/Evans/Smithies work dates to 1989 and the knockouts they made can now be ordered online. By the time the haploid human genome was announced to great fanfare in 2001, a lot of us thought that DNA was very old news. It may have been the "Book of Life" for geezers, but young folk viewed all those megabytes in the GenBank as just another online catalog.[11] Canonical runs of DNA were good enough for post-docs making photocopies, but for a kinetic, 3-D look at how genes work, the smart money was on RNA.

All that changed in the last two years. In the first place, Craig Venter added a surprising companion to the genome catalog: the Book of Life turned out to contain more typos and misprints than anyone predicted. Equally unexpected was the discovery that the dull routine of DNA replication has a novel twist: one strand of the double helix turns out to be "immortal," at least in stem cells.[12]

DIPLOID DNA

In September 2007, J. Craig Venter made public the DNA sequence of his own forty-six chromosomes. The original human genome sequences, reported in 2001—that Book of Life—relied on data from a haploid set of twenty-three chromosomes derived from a pool of donors. Venter's new diploid genome, derived from one paternal and one maternal set of chromosomes, suddenly made DNA much more interesting. Earlier genome sequencing projects, which attributed the bulk of human-to-human variations to SNPs (changes in single DNA bases), had stopped with sequences of about 13,000 bases. But now Venter's group was able to define runs of DNA hundreds of thousands of bases long. Comparing alternate alleles with those registered by the National Center for Biotechnology Information, Venter, et al. identified more than 4.1 million DNA variants. Encompassing 12.3 megabases, the variants yielded the surprising news that 44 percent of the genes Dr. Venter inherited from his mother were different from those that came from his father. A third of these variations had never been described and were by no means limited to SNPs.[13]

Venter's institute soon spread the news that human beings turn out to be much less alike than we ever suspected, at least five times less. Human-to-human variation, they proclaimed, is clearly greater than the 0.1 percent difference found in 2001. Their new estimate was that genomes vary between individuals by at least 0.5 percent.[13] The large number of variations has implications for genetic screening. In the near future, it seems likely that we'll be able to choose whether or not to know our own diploid genome. Should we? Variation gives hope to the genetically challenged, for if one parental strand has a spelling error, it is always possible that there is a spell-checker hidden in the other.

Venter told the press, "I might want to know: Do I have an additive risk from the genomes from both my parents, or did I get some helpful ones from her that counteract the ones from him?"[14] Others may not want to know: At a recent public event Charlie Rose asked Nobelist Joseph Goldstein of Dallas whether he was curious to know his genetic printout. Goldstein replied, "Look, if there is no more than a 15–20 percent concordance for colon cancer between identical twins who have 100 percent identical genomes, I'd get a colonoscopy."[15]

IMMORTAL COILS

"For in that sleep of death what dreams may come,
When we have shuffled off this mortal coil . . ."

—*Hamlet*

Perhaps it's not surprising that two parental genomes compete for our phenome—what we look like—we all have two parents. What's more astonishing is that the two complementary strands of DNA don't always uncoil equally when our cells divide. In 1975, John Cairns of Oxford floated the notion of "immortal DNA." He noted that our tissues contain both stem cells and differentiating cells. And since differentiating cells replicate quicker than stem cells, their DNA is a better target for the slings and errors of frequent replication. Stem cells, which replicate very slowly, would be likely to hold on to the original, error-free strands of DNA.[16] That seemed a reasonable explanation for why the most rapidly replicating cells in our body (e.g., in skin, gut, or mammary glands) are more likely to become malignant. In this process, now defined as "asymmetric self renewal,"[17] each adult stem cell undergoes a division that yields a new, pristine adult stem cell and its error-prone sister, which is the progenitor of the differentiated cells in the tissue. The mechanism of asymmetry is obscure.[18]

While Cairns's theory has led to lively and at times amusing debate, recent studies support the notion of "immortal DNA." Jim Sherley's laboratory at MIT was the first to show that individual mammalian cells in culture can undergo asymmetric self-renewal[19] and the process was soon documented in neural cells.[20] Further confirmation came a year ago from the Institut Pasteur. By following the fate of mouse satellite cells (muscle stem cells) in the course of cell division, the laboratory of Shahragim Tajbakhsh confirmed that stem cell DNA strands are distributed asymmetrically.[21] But the chemistry of DNA asymmetry remains as much of a puzzle in Paris as it was in Oxford. Tajbakhsh confessed, "How the cellular machinery distinguishes old DNA from new is still a mystery. For me, it is one of the most fascinating questions regarding DNA since the double helix was first described."[22]

Not to worry, *cher maitre*, I'd say. The mystery of asymmetry between mortal and immortal coils has been around a lot longer than the double helix.

THE CADUCEAN CHARM

The serpentine image of DNA has resonance with others in our collective history. From Apollo to Adam and Eve, Aesculapius to Moses, Hygea to Hermes, the coiled serpent has guarded mystery, knowledge, and healing. The caduceus, winged at the top and encoiled by twin serpents, is the symbol of Apollo's power and the eternal magic of snakes. The pythons who guarded Apollo's temple were believed to share with their master the divine arts of prophecy and healing, of prognosis and therapeutics. Aesculapius appeared in the dreams of his patients as an undulant snake, while Hygea was depicted with cup and adder. The caduceus was Apollo's gift to Hermes in trade for a stolen lyre, and Hermes used his wand to transmute species and to spell out their fate.

Nowadays, we've learned to play Hermes in the lab. Over fifty years ago Joshua Lederberg discovered homologous recombination in bacteria. Then Evans, Capecchi, and Smithies taught us how to recombine genes to engineer mice. Their Nobel citation explained, "It is now possible to introduce mutations that can be activated at specific time points, or in specific cells or organs, both during development and in the adult."[23] Transmutation and fate: what concordance between the myths of the ancient world and the latest news from Stockholm!

What we've also learned from knock-out and knock-in experiments is that if you know the genome, you can predict those "mutations that can be activated at specific time points" to affect you or your offspring. But, do we really want to know what mutations are entwined in our own diploid genome? Venter addressed those doubts in his *PLoS* paper:

> There are often concerns that individuals should not be informed of their predisposition (or fate) if there is nothing they can do about it. It is possible, however, that many of the concerns for predictive medical information will fall by the wayside as more prevention strategies, treatment options, and indeed cures become realistic. The cycle, in fact, should become self-propelling, and reasons to know will soon outweigh reasons to remain uninformed.[24]

KEATS'S *LAMIA*

The ever-smitten Hermes empty left
His golden throne, bent warm on amorous theft: . . .
For somewhere in that sacred island dwelt.
A nymph to whom all hoofed Satyrs knelt . . .

Lamia[25]

Nowhere in literature are Venter's "reasons to know" better defined than by John Keats's poem *Lamia*. These days *Lamia* can be read as a commentary on DNA itself: the split between mortal and immortal coils, the tricks of homologous recombination, and the penalty of remaining uninformed.

Lamia begins with the ever-smitten Hermes leaving golden Olympus to chase a nymph over hill and dale in Crete. The nymph becomes lost and Hermes is forlorn, but suddenly the messenger stumbles across a coiled creature,

. . . a palpitating snake,
Bright, and cirque-couchant in a dusky brake. . . .
Her head was serpent, but ah, bitter-sweet!
She had a woman's mouth with all its pearls complete;[26]

The reptile, a hybrid of mortal and immortal strands, exacts a promise from Hermes to transmute her into human form. In return for the gift of recombination, Lamia will tell Hermes where his nymph is hidden. It all works as promised: the nymph is found, Hermes exults, and hybrid Lamia swoons. Fulfilling his part of the bargain, Hermes turns with snake-entwined wand

To the swoon'd serpent, and with languid arm,
Delicate, put to proof the lithe Caducean charm. . . .
Left to herself, the serpent now began
To change; her elfin blood in madness ran;

What a Romantic precedent for homologous recombination and rapid differentiation! But Lamia is more than just a pretty transgenic face. Since she was constructed to retain and express on induction the serpentine genes of passion, she became

A virgin purest lipp'd, yet in the lore
Of love deep learned to the red heart's core:

Thus equipped, Lamia hies to Corinth, where she ensnares a young philosopher (read, scientist), Lycius. The two become enraptured with each other, but Lycius deliberately ignores the different worlds they have inhabited. Soon enough, they exchange vows of wedlock.

But at a drunken prenuptial feast, their fate is sealed. One of the wedding guests is Apollonius, a sophist, who has been Lycius's mentor. This "bald-head philosopher" (read, thesis advisor) instantly spots Lamia as a dangerous demon and fixes her in his withering gaze. When Apollonius denounces Lamia as a dangerous serpent, the beauty blanches. Lamia turns white, then cold, then vanishes into thin air. She has reverted to the demon world.

Lycius cannot bear this loss; he dies a languorous death of grief, having paid the final penalty for remaining willfully uninformed. The tale could be read as an augury of Venter's prediction that the reasons to know will soon outweigh the reasons to remain uninformed.

Keats, like many Romantics, worried that the Newtonians of the Royal Society had destroyed the beauty of the rainbow itself.

> . . . Do not all charms fly
> At the mere touch of cold philosophy?
> There was an awful rainbow once in heaven:
> We know her woof, her texture; she is given
> In the dull catalogue of common things.
> Philosophy will clip an Angel's wings . . .
> Unweave a rainbow, as it erewhile made
> The tender-person'd Lamia melt into a shade.[27]

Keats was afraid that exact knowledge of material nature, the dull catalog of common things, would destroy not only aesthetics but ethics as well. These days we'd call that dull catalog of common things our diploid genome (± 0.5%), and hope that our ethics can cope with its challenges.

I find Venter's peroration in *PLoS* reassuring:

> *Ultimately, as more entire genome sequences and their associated personal characteristics become available, they will facilitate a new era of research into the basis of individuality. The opportunity for a better understanding of the complex interactions among genes, and between these genes and their host's personal environment will be possible using these datasets composed of many genomes. Eventually, there may be true insight into the relationships between nature and nurture, and the individual will then benefit from the contributions of the community as a whole.[28]*

2. Homeopathy: Holmes, Hogwarts, and the Prince of Wales

Oliver Wendell Holmes, M.D. (1880)

If you had a bent tube, one arm of which was of the size of a pipe-stem, and the other big enough to hold the ocean, water would stand at the same height in one as in the other. Controversy equalizes fools and wise men in the same way,—and the fools know it.

—O. W. HOLMES, *The Autocrat of the Breakfast-Table*[1]

I just come and talk to the plants . . . and they respond.

—PRINCE CHARLES, August 29, 2007, *Daily Mail*

T HE HYDROSTATIC PARADOX HAS NEVER BEEN SO WELL ILLUSTRATED AS BY current discussions of alternative medicine and its poster child, homeopathy. Hahnemann's system, a therapeutic regimen unchanged since the age of Mesmer, is making a comeback in the age of Andrew Weil. In 1810, Hahnemann (1755–1843) rebuked Enlightenment medicine in an over-ideational treatise called *The Organon of the Rational Art of Healing*:

The partisans of the old school of medicine flattered themselves that they could justly claim for it alone the title of "rational medicine," because they alone sought for and strove to remove the cause of disease . . . [but] the greatest number of diseases are of dynamic (spiritual) origin and dynamic (spiritual) nature, their cause is therefore not perceptible to the senses.[2]

Not content to play the spiritual card, Hahnemann took swipes at the science of his day. Anatomy, physiology, and pathology, he argued, presented only "dim pictures of the imagination." Since disease was not caused by any discrete physical agent, but by man's lack of harmony with the "vital force" of nature, he asked, "Has any one ever succeeded in displaying to view the matter of gout or the poison of scrofula?"[3]

More than a century after crystals of monosodium urate were shown by Garrod to be the matter of gout, and the poison of scrofula was found by Koch to be M.tuberculosis, homeopaths still believe that *Organon*'s vital force of nature is at the root of gout and TB.[4]

A PRINCE AT WAR WITH SCIENCE

In May 2006, Prince Charles addressed the World Health Assembly in Geneva to argue for homeopathy and its kindred therapies. He urged a return to remedies "rooted in ancient traditions that intuitively understood the need to maintain balance and harmony with our minds, bodies, and the natural world." He complained about modern biomedicine: "It seems to be that in our ceaseless rush to modernize, many tried and tested methods which have shown themselves be effective have been cast aside as old-fashioned or irrelevant to today's needs." News flashed around the world: "CHARLES AT WAR WITH DOCTORS."[5] The Prince of Wales, whose oddball habit of "talking to plants" is widely mocked,[6] has been at war with medical science for some time. In 1985 he caused a stir by warning the British Medical Association that "the whole imposing edifice of modern medicine, for all its breathtaking successes is, like the celebrated Tower of Pisa, slightly off balance."[7] Last year, the Prince funded a commission headed by a bank executive as lacking in scientific credentials as Charles himself, to "look at the effectiveness, especially from a financial point of view, of integrated healthcare."[8] The commission claimed that up to 480 million pounds could be cut from the prescription drugs bill of Britain's National Health Service if 10 percent of primary care physicians offered homeopathy as an alternative to standard drugs. Of course, had 20 percent of the docs offered rose water to their patients, Britain could have saved a billion pounds.

British science struck back. Anticipating Prince Charles's sermon in Geneva, thirteen of Britain's most eminent physicians and scientists issued a widely quoted open letter: "Re Use of 'Alternative' Medicine in the NHS."[9] The letter expressed concern over "ways in which unproven or disproved treatments are being encouraged for general use in the NHS [Britain's National Health Service]." The signatories, who included three Fellows of the Royal Society, one Nobel Laureate (Sir James Black, FRS), and the son of another (Professor

Gustav Born, FRS), cited the overt promotion of homeopathy by the NHS, including its official website. The open letter warned that "it would be highly irresponsible to embrace any medicine as though it were a matter of principle."

Their position was supported by an extensive meta-analysis in *The Lancet* of the efficacy of homeopathy which documented that homeopathic regimens were no better than a placebo for a wide variety of ailments.[10] The open letter also concluded that homeopathy "is an implausible treatment for which over a dozen systematic reviews have failed to produce convincing evidence of effectiveness." They should know: Sir James's highly effective beta-blockers and H2 antagonists have kept more humans alive than any integrated crystal therapist, and if Gustav Born hadn't worked out platelet aggregation, we'd have missed the aspirin effect in heart disease.

As for the Prince's "financial point of view," Professor Michael Baum, Britain's leading breast cancer authority, noted that Britain recently spent twenty million pounds refurbishing the Royal Homeopathic Hospital. Had that sum of money been spent on making available herceptin and aromatase inhibitors for breast cancer victims, it could have saved 600 lives a year in one health district alone.[11]

Prince Charles was unfazed; on the day the open letter was published, he stopped at St. Tydfil's Hospital in South Wales to watch alternative medicine at work. He accepted a "spiritual" crystal, as if he were Albus Dumbledore, headmaster of Hogwarts School, consulting the Wizard's sphere. Unlike Dumbledore, however, who only professed Witchcraft and Wizardry, Prince Charles called up every form of "integrative therapy" against Alzheimer's disease.[12] One notes that when Prince Charles and other fans of unproven or disproved medical practices use terms such as "integrated therapy" or "alternative medicine," they're following the lead of creationists who hide under the term "intelligent design." These are all convenient slogans that permit the credulous to con the gullible.

HOGWARTS IN BETHESDA

The Prince's activities have not been limited to the UK. In 2003, he authorized his U.S. charity to fund a research fellowship at NIH's National Center for Complementary and Alternative Medicine (NCCAM). This unusual gift to an agency of the U.S. government came after a private dinner at St. James's Palace to which the Prince and Camilla Parker-Bowles had invited the clinical director of NCCAM, Marc Blackman, his wife, and like-minded guests "to discuss ideas and visions for complementary medicine."[13]

The Center has expended much of its 120-odd-million-dollar war chest to fund studies on chelation therapy, black cohosh, mushroom-induced immunopotentiation, and homeopathic dilutions of cadmium for prostate cancer. It recognizes and explains in detail therapies based on "the life force" variously called *"qi," "ki," "dosha," "prana,"* "etheric energy," *"fohat,"* "orgone," "odic force," *"mana,"* and "homeopathic resonance."[14]

The NIH seems happy with research on homeopathy and kindred therapies.[15] Its website replies yes to the question "Is NCCAM funding research on homeopathy?" while admitting that "homeopathy is an area of complementary and alternative medicine (CAM) that has seen high levels of controversy and debate, largely because a number of its key concepts do not follow the laws of science (particularly chemistry and physics)." Never mind the laws of chemistry and physics: this sounds like a final exam at Hogwarts!

However, when it comes to homeopathy, NCCAM is careful to issue a disclaimer: "It has been questioned whether a remedy with a very tiny amount (perhaps not even one molecule) of active ingredient could have a biological effect, beneficial or otherwise." Nevertheless, NCCAM has contributed $250,000 toward a clinical trial of "verum LM" (a homeopathic medicine diluted 1:50,000) for fibromyalgia at Dr. Andrew Weil's Program in Integrative Medicine at the University of Arizona. Weil's the guru who in a 1986 book, *Health and Healing*, announced that "sickness is the manifestation of evil in the body."[16] Hahnemann redux, one might say. Since many rheumatologists believe that "fibromyalgia" is simply a euphemism for a variety of medicalized miseries,[17] it was not unexpected that Arizona's integrators found useful a remedy that may or may not contain one molecule of active ingredient for a disease that may or may not exist.[18]

DILUTIONS OF GRANDEUR

In 1834, Dr. Oliver Wendell Holmes, poet, essayist and Professor-to-be of Anatomy at the Harvard Medical School, returned to Boston from study in Paris with microscope in hand. He was convinced that French quantitative science would find a new home in the city, which he christened "the Hub" of the universe. But Boston in 1834 had permitted curious forms of the healing art to flower and Holmes was appalled. By 1842, he'd had enough and wrote the definitive critique of the practice: "Homeopathy and Its Kindred Delusions."

> *It takes a very moderate amount of erudition to unearth a charlatan like the supposed father of the infinitesimal dosing system. . . . [But] Homeopathy has proved lucrative, and so long as it continues to be so will surely exist,—as surely as astrology, palmistry and other methods of getting a living out of the weakness and credulity of mankind and womankind.[19]*

Holmes's arguments hold today and are worth re-examining. Hahnemann's *Organon* lays down three proposals that have become laws to his followers: the "Law of Similars," the "Principle of Minimum Dose," and "Prescription for the Individual."[20] Now, individual prescription seems harmless enough and one can give Hahnemann a pass; the other two are worthy of Hogwarts.

The Law of Similars states "the principle that like shall be cured by like, or *similia similibus curantur*." If a substance produces symptoms of illness in a well person when administered in large doses, administered in minute quantities, it will cure disease in a sick person. Nonsense: arsenic poisoning produces bloody diarrhea; minute amounts of arsenic (as in the drinking water of Bangladesh) don't prevent shigellosis.

NCCAM demurs: "It is debated [*sic*] how something that causes illness might also cure it."[21] Yes, but only in the sense that there is a scientific "debate" between those who hold the theory of evolution and those who believe in intelligent design. NCCAM may be said to be "teaching the controversy" between the laws of chemistry and physics and the cult of the *Organon*.

Now for the Principle of Minimum Dose: it turns out that most homeopathic solutions contain nothing at all. Sad to say, the last remaining privately owned drugstore in my neighborhood features dilutions of Oscillococcinum®, a "200C" product. That's a dilution number, cunningly calculated to guarantee that the original ingredient has been diluted several million times, and has surely exceeded Avogadro's number. The magic numbers have been calculated by Dr. Stephen Barret as follows: "Dilutions of 1 to 10 are designated by the Roman numeral X (1X = 1/10, 2X = 1/100, 3X = 1/1,000, 6X = 1/1,000,000). Similarly, dilutions of 1 to 100 are designated by the Roman numeral C (1C = 1/100, 2C = 1/10,000, 3C = 1/1,000,000, and so on). Most remedies today range from 6X to 30C, but some carry designations as high as 200C. Oscillococcinum®, that 200C product "for the relief of colds and flu-like symptoms," involves dilutions that are even more far-fetched. Its "active ingredient" is prepared by incubating small amounts of a freshly killed duck's liver and heart for 40 days.[22]

Were a single molecule of the duck's heart or liver to survive the dilution, its concentration would be 100^{200}. This huge number, which has 400 zeroes, is vastly greater than the estimated number of molecules in the universe (about one googol, which is a 1 followed by 100 zeroes). Quackwatch—a website well named for a duck authority—quotes the February 17, 1997, issue of *U.S. News & World Report* as reporting that only one duck per year is needed to manufacture the product, which had total sales of $20 million in 1996. The magazine dubbed that unlucky bird "the $20-million duck."[23] But of course it could be argued that at least these dosages do no harm. I'd remind you of the legendary patient who was mistreated by a homeopath and nearly died of a massive underdose.

HONORS FOR HOGWARTS

It's been said that "if homeopathy here in the United States has been a leader at all, it has to be Dana Ullman!"[24] Mr. Ullman has written six major texts, and Penguin, his publisher, boasts that he serves on the Advisory Council of the Alternative Medicine Center at Columbia University's College of Physicians and Surgeons and is a consultant to Harvard Medical School's Center to Assess Alternative Therapy for Chronic Illness.[25]

Penguin does not report that in the course of the anthrax outbreak in October 2001, Mr. Ullman advised use of the homeopathic medicine Anthracinum for the prevention and treatment of anthrax. This agent, gathered from infected swine, is called a nosode and its producers reassure the public that they are "diluted to a point where no molecules of the disease product remain."[26] Well, nosodes and Anthracinum, miasmas and the like, which dot the Hogwarts curriculum of Mr. Ullman's site, are matched by those on the websites of the university centers he has advised. Columbia's Rosenthal Center offers "integrative medicine" for children with cancer offered by a staff experienced in *Shiatsu*, reflexology, aromatherapy, *Reiki*, Flow Alignment and Connection, *So Tai*, and *Tui Na*.[27] Not to be outdone, Harvard Medical School's Osher Institute offers clinical fellowships, funded by NCCAM, to study remedies that meet Prince Charles's criterion of being "rooted in ancient traditions": acupuncture, herbal therapies, chiropractic, relaxation techniques, therapeutic massage and other proto-scientific measures that sidestep the laws of chemistry and physics.[28] It is in this context that one can understand why, when Columbia's Rosenthal Center kicked off its tenth anniversary celebration on November 20, 2003, at the St. Regis Hotel in New York, the awardees of honor were—you guessed it— Dr. Andrew Weil and Prince Charles.[29] Is Albus Dumbledore next?

Hogwarts is certainly on the move! If the trend persists, perhaps MIT or Cal Tech will march in step with the medical schools and offer prizes for alternative and complementary alchemy or alternative aeronautics. But Dr. Oliver Wendell Holmes, dean of the Harvard Medical School before the age of Oprah, had the last word on homeopathy:

> *Some of you will probably be more or less troubled by the pretensions of that parody of mediaeval theology which finds its dogma in the doctrine of homeopathy—its miracle of transubstantiation in the mystery of its . . . dilutions, its church in the people who have mistaken their century, and its priests in those who have mistaken their calling.*[30]

3. Citizen Pinel and the Madman at Bellevue

Tony Robert-Fleury (1838–1911), painting, 1876.
Citizen Pinel orders removal of the chains of the mad at the Salpêtrière.

David Tarloff, 39, remained at Bellevue Hospital Center, where he was under evaluation after his arrest in the killing of a psychologist, Kathryn Faughey, on the Upper East Side last week. . . . "He didn't really say anything," [his father] said. "He didn't talk about the situation. He's in a fog." —New York Times, *February 20, 2008*[1]

Among madmen . . . there are some whose imaginations are by no means affected, but who feel a blind and ferocious propensity to imbue their hands in human blood.
—PHILIPPE PINEL *(1806)*[2]

In the hall in which [Charcot] gave his lectures hung a picture of the "citizen" Pinel causing the poor insane of the Salpêtrière to be relieved of their chains; for after having been the scene of so many horrors during the French revolution, the Salpêtrière had also witnessed this most humane innovation. —SIGMUND FREUD *(1893)*[3]

A FEROCIOUS PROPENSITY

AFTER HIS INDICTMENT FOR MURDER IN FEBRUARY 2008, DAVID TARLOFF was kept on a locked ward three floors above my old lab at Bellevue Hospital. He remained accused of murdering a psychologist, Kathryn Faughey, with a meat cleaver in her office. The case attracted wide attention not only for its ferocity, but also for illustrating how antipsychotic drugs have changed the way we deal with the demented. The *New York Post* demanded prompt action against LETTING MADMEN ROAM,[4] while the *New York Daily News* called for preemptive strikes by either our governor or mayor: ELIOT, MIKE; YOU MUST ACT.[5] The *News* worried that Tarloff had "passed through New York's mental health system like water in a sieve." It asked, "Will such madness never end? It must. Governor Spitzer and Mayor Bloomberg must act now."[6] A civic newsletter asked, "IS MURDER THE THRESHOLD?" for involuntary detention and urged that "judges and district attorneys should be educated as to the potential danger posed by violent mental patients."[7]

Tarloff, diagnosed as a paranoid schizophrenic in 1991, has a long psychiatric history. His family repeatedly tried to have him committed; yet he bounced in and out of a dozen institutions in the area. He'd also been treated on and off with lithium, Depakote®, Haldol®, Seroquel®, and Zyprexa®—agents as useful in mood disorders as in schizophrenia. One week before the murder, he was arrested for violent behavior at a community hospital. An uncertified practitioner found him lucid, and Tarloff was dismissed with a slap on the wrist by Judge Barry Kron:

> KRON: *"The defendant is 40 years old, has no prior record, is under stressful circumstances, and anybody who has ever been in a hospital dealing with the personnel there can appreciate the frustration one deals with. You need to use restraint and common sense, and if there's any sort of problem . . . you never want to come back here again."*
>
> TARLOFF: *"I'm sorry."*[8]

After this—by then routine—encounter with the legal system, a chastened, unmedicated David Tarloff was released into "the community" to work out his demons by slaughtering Kathryn Faughey,

After his arrest and arraignment, at which he "babbled aimlessly," Tarloff was remanded to Bellevue for medication and evaluation. Next day, two of our staff psychiatrists found him mentally fit to stand trial, and, as expected, Tarloff's own lawyer demanded an independent examination. While the criminal justice and health care systems debated his future, Tarloff was remanded without bail. Those who saw him at Bellevue described him as heavily sedated and essentially docile.[9] No wonder. He was under restraints as effective as the straitjackets or shackles of yesteryear: antipsychotic drugs.

PEOPLE WHO CAN'T FIT IN

These days the topsy-turvy care of schizophrenic patients is the resultant of several vectors: our social attitudes toward mental illness in general; our rudimentary understanding of the biology of schizophrenia; and, since Thorazine, the prescription of psychotropic drugs taken, or not taken, by the patient. Unfortunately the social urges that emptied our mental hospitals and asylums have outlived their day. The therapeutic rescue fantasy of the 1960s was that the mentally ill would be freed of their symptoms by medicinal chemistry and returned to the community. Decades of clinical experience have dashed those hopes. Dr. Carol Tamminga, an editor of the *American Journal of Psychiatry*, admits, "People say there are drugs to treat schizophrenia. In fact, the treatment for schizophrenia is at best partial and inadequate. You have a cadre of cognitively impaired people who can't fit in."[10]

It does seem a bit far-fetched to believe that the social structure of our large cities bears any resemblance to the nurturing "community" that might support the fragile psyche of the mentally ill. Cast into an environment limited by the overnight shelter and the park bench, lacking adequate monitoring of their medication, they fall prey to climatic extremes and urban predators. Many deinstitutionalized schizophrenics wind up as conscripts in a vast army of the homeless quartered in temporary shelters such as church basements, armories, and flophouses. In Irvine, California, unused animal shelters were opened to the vagrant.[11]

Lost in the legions of the homeless, a good number of partially treated, seriously deranged schizophrenics become involved in violent crime, either as subject or object. Emanuel Tanay, a forensic psychiatrist at Wayne State, sounded a warning that resonates at Bellevue:

> *The failure to hospitalize potentially violent schizophrenics contributes to the incidence of psychotic homicide [and has] brought about a shift of this population from the mental health system to the criminal justice system.*[12]

Not just in the U.S. Austrian doctors have found that the odds of finding a male schizophrenic among murderers is 5 times that in the general population.[13] In Finland, where ascertainment is even better, since "police solve 95% of all homicides," the odds are 10 to one.[14] Worse news yet: released male forensic psychiatric patients, mainly schizophrenics, were 300 times more likely than the general male population to commit another murder in the year after discharge.[15] Psychotic homicide, of course, is only one side of the equation. These poor, deranged folk are not only ticking time bombs; they are also the walking wounded who themselves are open to violence in the asphalt jungles of our cities.[16]

One could summarize the experience of two generations in our treatment of schizophrenia: In the 1950s, the sick were warehoused in heated public hospitals with occasional visits by trained psychiatrists who dispensed pills that didn't work. From the 1960s to the 1990s, the medicines worked better, and patients wandered in the cold without access to psychiatrists to check their pills. We've made progress, however. Today, when the streets become too cold, we warehouse the mentally ill in crowded shelters or jail them when they commit a crime. Schizophrenics have access to trained professionals again, but to cops, not psychiatrists.

THE RECEPTION OF MADMEN

Clearly, we need to re-create a new asylum: a space in which the mad and deranged can be protected from their manias and the brutality of the street. As usual, we could do worse than to consult the Age of Reason for a model on which to build.

In 1792, at the men's asylum of Bicêtre prison, Dr. Philippe Pinel removed the shackles from the limbs of mental patients and introduced mental hospitals to Enlightenment thought:

> *In lunatic hospitals, as in despotic governments, it is no doubt possible to maintain, by unlimited confinement and barbarous treatment, the appearance of order and loyalty. The stillness of the grave, and the silence of death, however, are not to be expected in a residence consecrated to the reception of madmen.*[17]

Philippe Pinel (1745–1826) had been appointed to the Bicêtre by the Revolutionary authorities in 1793 and soon put into action the philanthropic principles of the *philosophes*. His unshackling of the mad was of a piece with Denis Diderot's campaign for the humanity of blind folk, and Abbé Gregoire's plea for equality between Whites and Negroes, Jew and Gentile. Pinel was also cast in the mold of such eighteenth-century taxonomists and encyclopedists of nature as Linnaeus and Buffon: but the atlas of natural history he assembled was one of mental disease. In his 1801 "Treatise on Insanity"[18] Pinel documented the number of patients admitted to his care at Bicêtre, their ages, diagnoses, length of stay, the temperature of patients and of ambient air, the outcomes of baths and showers, of bleedings or purgatives, the number who died, who recovered, etc. The mad were classified according to his own *Nosographie* of 1789, which sorted mental disease into five categories: melancholia, mania without delirium, mania with delirium, dementia, and idiocy.[19] Pinel would be pleased to consult today's psychiatry texts, or even the *DSM-IV*, to find that not much has changed in the nosology game.[20]

Pinel was also an experimentalist whose careful measurements undermined the phrenology of his day. Using an instrument based on the parallelopipedon (a prism whose bases are parallelograms), he made painstaking measurements of dozens of human skulls, of madmen and -women, of patients with and without delirium. He contrasted these with the dimensions of the skulls of congenital idiots, and concluded that the fault in delirium lay not in our skulls, but inside.

> *The anatomical examination . . . would appear to confirm the opinion which I have already advanced, that intense mental affections are the most ordinary causes of insanity, and that the heads of maniacs are not characterised by any peculiarity of conformation that are not to be met with in other heads taken indiscriminately.*[21]

It has taken modern neuroimaging, using CT, MRI, PET, SPECT, fMRI, and DTI to identify consistent neuroanatomical findings that correlate with those "intense mental affections" found in schizophrenia.[22]

But the mythical Pinel is not the remote taxonomist of madness or the anthropologist of skulls. He's the citizen-hero of the painting in the Salpêtrière that Freud saw many years later.[23] Tony Robert-Fleury's painting turned a historical incident in the all-male Bicêtre of 1793 into a saltier version at the women's asylum of the Salpêtrière, sometime after 1795. As reported by Pinel's son, his father had been confronted by Georges Couthon, an official of the Commune in its Reign of Terror phase. Couthon was seeking "hidden traitors" to bring to trial. Pinel led him to the cells of the most seriously disturbed, where Citizen Couthon's attempts at interrogation were greeted with "disjointed insults and loud obscenities." It was useless to prolong the interviews. Couthon turned to Pinel and asked:

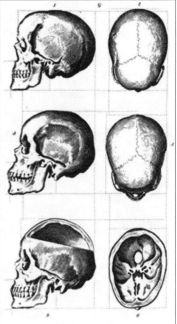

> *"Now, Citizen, are not you mad yourself to think of unchaining such animals?" Pinel replied: 'Citizen, I am convinced that these lunatics have become so unmanageable solely because they have been deprived of air and liberty. . . .'"*

> *"Well, then do as you like with them, I give them up to you. But I fear you may become victim of your own presumption."*[24]

The official left, and Pinel removed the chains. He freed twelve of the most violent as an experiment, but as a safeguard, had prepared twelve "long-sleeved waistcoats fastened at the back" (straitjackets), in case the freed lunatics proved unmanageable. His plan worked, and when Pinel moved his operation from the Bicêtre to the Salpêtrière in 1795, his novel mix-

Philippe Pinel, *Projections of the normal skull.* From *A Treatise on Insanity* (1806).

ture of carrot and stick went with him. He launched the asylum model of firm but supportive care, a model that had the Rights of Man as its base.

A degree of liberty, sufficient to maintain order, dictated not by weak but enlightened humanity, and calculated to spread a few charms over the unhappy existence of maniacs, contributes, in most instances, to diminish the violence of the symptoms, and in some, to remove the complaint altogether.[25]

A PRISONER OF NOTHING BUT HIMSELF

Our loss of the Enlightenment asylum as central to the treatment of the severely disturbed can be traced directly to the advent of first- and second-generation psychotropic drugs which by now act on every known class or subclass of neuroreceptor—histamine, dopamine, glutamine, serotonin, GABA, etc.[26, 27] But the release of inmates would not have become widespread were it not for activists convinced that, in an age of the gulag, involuntary commitment of the mad might lead to Soviet-style incarceration for dissidents.

Another force in the attack on the asylum was led by the late Michel Foucault, a once-trendy French historian of sexuality, prisons, and medicine. Nowadays his polemics seem even less convincing than Timothy Leary's odes to LSD, but at the time they roused a generation of young social scientists to arms. In his major opus, *Madness and Civilization,* Foucault traced the history of mental institutions to the "Great Confinement" of the Enlightenment years. "The asylum was substituted for the lazar house," he wrote, referring to the leprosariums of the Middle Ages:

[T]he old rites of excommunication were revived . . . and the nineteenth century would consent, would even insist that to the mad and to them alone be transferred these lands on which, a hundred and fifty years before, men had sought to pen the poor, the vagabond, the unemployed.[28]

One would have thought that Foucault would have been a great champion of Pinel's brand of philanthropy. But, no. In keeping with a dated tradition of rhetoric associated with the names of Herbert Marcuse and Ivan Illich, Foucault reserved his hardest blows for Pinel and Enlightenment thought. Foucault accused Pinel of substituting psychological chains, the "moral treatment of the mad," for the iron shackles of the old lazar house. Foucault, displaying that curious affinity of advanced French thought for the punitive tableaux of the Marquis de Sade, went on to grieve for the patients unchained after Pinel's reform. He posed a very fashionable Gallic paradox:

The dungeons, the chains, the continual spectacle, the sarcasms were, to the sufferer in his delirium, the very element of his liberation. . . . But the chains that

fell, the indifference and silence of all those around him confined him in the lim-
ited use of an empty liberty. . . . Henceforth, more genuinely confined than he
could have been in a dungeon and chains, a prisoner of nothing but himself, the
sufferer was caught in a relation to himself that was of the order of transgression,
and in a non-relation to others that was of the same order of shame.[29]

THE OTHER DEPARTMENTS
OF NATURAL HISTORY

In each critique of the asylum, Foucault used as his point of reference a
golden age of madness, a medieval fantasy of monk and mountebank, where
fools and crazies added to the richness of everyday life by their unique insights
and startling behavior. The "reforms" of the Age of Reason were said to have
destroyed this organic fabric and turned it into Pinel's straitjacket. According
to this popular storyline, psychiatrists, psychologists, and social workers were
employees of a police state designed to censor the self-expression of the mad.
Those accusations of the 1960s were accepted by well-meaning intellectuals of
the West and entered the canon of modern social science. They also caught
the ear of every municipal, state, and federal official who ever had a budget to
balance. In consequence, for two generations now, the asylums have been
emptied, mental health budgets cut, and church basements filled.

In American cities, where violent schizophrenics wander the streets, where
madmen readily buy attack rifles, where aggressive thugs seek drugs and easy
victims, our streets and malls revert at times to the Hobbesian state of nature.
But trend is not destiny, as Lewis Mumford taught, and therefore I, too, have
a vision of a golden age when diseases of the mind will have yielded to exact
knowledge. From neuroimaging, we'll have learned whether the dopaminer-
gic or glutamatergic hypotheses reflect primary defects; molecular genetics
will have cleared up the 50 percent concordance rate in identical twins; and
we'll have discovered interventions that really work.[30] Thanks to our new natural science, the schizophrenias will have come to be as well understood and
treated as scarlet fever or rabies. Those diseases were intractable on the wards
of the Salpêtrière in 1795, but yielded their secrets to Pasteur when French science itself was finally unfettered. The rabid are no longer restrained.

Citizen Pinel hoped that

The time, perhaps, is at length arrived when medicine in France, now liberated
from the fetters imposed upon it, by the prejudices of custom, by interested ambi-
tion, by its association with religious institutions, . . . will be able to assume its
proper dignity, to establish its theories on facts alone, to generalise those facts, and
to maintain its level with the other departments of natural history.[31]

4. The Experimental Pathology of Stress: Hans Selye to Paris Hilton

Above, Paris Hilton (1981–)
Left, Hans Selye (1907–1982)

Stress is the nonspecific response of the body to any demand. A stressor is an agent that produces stress at any time. The general adaptation syndrome (GAS) represents the chronologic development of the response to stressors when their activation is prolonged. It consists of three phases: the alarm reaction, the stage of resistance, and the state of exhaustion. —HANS SELYE, "Forty Years of Stress Research" [1]

Paris Hilton has stopped eating under the stress of her impending jail sentence.
—"Paris Hilton's Stress Starvation," *Boston Globe* [2]

> *She walked among the Trial Men*
> *In a suit made by Versace . . .*
> *But I never saw a girl who looked*
> *So fond of paparazzi.*
> *The Warden said that Law was Law*
> *And Paris no exception*
> *And all the day she phoned her friends*
> *And missed their soft caress*
> *And all her faithful flock proclaimed*
> *"She must be under stress"*
> —"Verses on the incarceration of Paris Hilton," D. J. Taylor
> *(after Oscar Wilde).* The Independent, *June 2007* [3]

Sergeant Padilla, 28, could not ward off memories of the people he had killed with a machine gun perched on his Bradley fighting vehicle. On April 1, according to the authorities and friends, he withdrew to the shadows of his Colorado Springs home, pressed the muzzle of his Glock pistol to his temple and squeezed the trigger. Sergeant Padilla had been diagnosed with post-traumatic stress disorder at Fort Carson Army base.
—New York Times, *May 13, 2007* [4]

THE LIBERATION OF PARIS

T HE DAY AFTER PARIS HILTON WAS RELEASED FROM THE LOS ANGELES County jail she confessed to Larry King on CNN that "I've been through a lot and it was a pretty traumatic experience." Recounting traumas that included bad food and nightmares, the tabloid princess told millions of her subjects that "I just want to let people know what I went through."[5] Even before Ms. Hilton was jailed for violating a court order, she had stopped eating "under the stress" of her impending stay in the hoosegow. Since her stress preceded the trauma, Paris Hilton may be the first well-documented case of "Pre-Traumatic Stress Disorder." Not to worry though; after twenty-three days in the jug, the heiress was whisked away to the family mansion, followed by brigades of paparazzi. The shutterbugs were stopped at the gates of the compound, which were then opened wide for a balloon-decorated SUV bearing a "Welcome Home" cake in pink frosting and a van from Dream Catchers Hair Extensions.[6] The Dream Catchers craft was much in evidence on CNN as Ms. Hilton recounted the ritual of celebrity stress and redemption. In keeping with the customs of the Larry Kingdom, she owned that she had found her "spirituality" in prison and that she will continue to confide in her therapist: "I talk with someone about my problems." As had Britney, and Anna Nicole, and Lindsay before her, she committed herself to rehab, or counseling—and God.[7]

It's a different story entirely for those who suffer from the far heavier burden of post-traumatic stress disorder (PTSD). Almost two out of every ten U.S. combat troops who return from action in Iraq show serious symptoms of depression, anxiety, or post-traumatic stress disorder. Sad to say, less than 40 percent of those afflicted by PTSD will seek help for *their* recurrent nightmares.[8] According to congressional testimony, suicide rates due to PTSD after service in Iraq have mounted, while VA facilities for psychiatric care have proven inadequate.[9,10] This treatment gap carries a toll: EX-GI DIAGNOSED WITH PTSD DIES IN 2-CAR COLLISION ON I-25 headlined the *Denver Post* last year.

> *Jessica Rich, who was driving the wrong way, couldn't shake some of the memories of the Iraq war, including witnessing a suicide. Rich, 24, who served in the Army Reserve, was killed at 10:25 p.m. Thursday when her 1996 Volkswagen Jetta smashed head-on into a 2003 Chevy Suburban. . . . Makayla Crenshaw, 24, a friend who had known Rich for six years and was in the same unit when they served in Iraq, said that when she talked to her friend two months ago, Rich said she was waiting to enter a treatment facility for PTSD.*[11]

Mental science has made some progress in defining the stresses of wartime trauma. The diagnosis of PTSD has been rigorously defined and quantified in the psychiatric literature,[12] and attention has now turned to its roots in basic biology.[13] The same cannot be said for the aches and pains in Larry Kingdom.

TOP BANANA

Ever since Hans Selye transformed "stress" into a force of nature called "STRESS" (his caps) over half a century ago,[14] the term has been applied to almost every sling and arrow endured by sentient creatures. At the Paris Hilton level, stress results from paparazzi at the door or hair in need of extension. It's perhaps too easy to poke fun at these talk-show notions of stress and their ramifications. But it's no laughing matter: Angela Patmore's recent *The Truth About Stress* estimates that the stress-management industry in the U.S. eats up about $18 billion each year.[15] Since the American Institute of Stress, which Hans Selye inspired, defines stress loosely as "The rate of wear and tear on the body,"[16] it's no wonder that "stress" has been tossed into the word-salad of psychobabble with such other ingredients as "anger-management," "self-esteem," "insecurity," and "closure." But stress is top banana: the institute claims that stress is "America's #1 Health Problem." In the 1940s Selye had formulated the notion that we owe each of our "specific" ills to a habitual, nonspecific response to stress in general;[16] by the 1950s stress was blamed for bringing on dropsy and scurvy, herpes and whooping cough, cancer and the common cold.[17] In *Guys and Dolls* (1950) Frank Loesser gave us Adelaide, an "unmarried, female, basically insecure," who develops the sniffles because she can't get hitched:

> In other words, just from waiting around for that plain little band of gold
> A person can develop a cold.[18]

Colds for Loesser, cancer for others. Woody Allen explains to Diane Keaton in *Manhattan* (1979):

> I don't get angry, okay? I mean, I have a tendency to internalize. I can't express anger . . . I grow a tumor instead.[19]

We read in Patmore's book that in the 1990s the World Health Organization called the stress of everyday life "a worldwide epidemic." Twenty-nine percent of Americans admit to having "yelled" because of stress, while 26 percent have been "driven to eat chocolate." Stress at this level is truly international: British stress-management gurus offer dolphin click noises, aroma pillows, and "squeezy water knobbled key-rings"; the country has a National Stress Awareness Day and offers a "stress bus tour" of central London.[20] In Spain, on July 4, 2007, a hotel planning renovations in Madrid offered thirty "highly stressed out people" selected by a team of psychologists the chance to take up sledgehammers and battering rams and smash through its rooms.[21] In France, a tour operator offers to "Reduce Stress!" by having clients choose the "most reliable and high quality supplier" of a

> *2-hour World War II Walk* [that] *will take you back to Paris' darkest hours—the Nazi occupation. Explore subjects such as the invasion of France, the Resistance, life in Paris during the occupation, the D-Day invasion and of course the Liberation itself.*

Ironically, this stress-reduced *recherche du temps perdu* will take tourists to

> *Hotel Meurice, the hotel that housed the Nazi headquarters during the occupation that still has a bullet hole in the door crest.*[22]

That stress-free walk by the Hotel Meurice recalls the Liberation of Paris (France, not Hilton). The Nazi commanders at the Hotel Meurice had turned over their most valuable captives to the Gestapo for torture at 84 Avenue Foch, the Abu Ghraib of its day. When the Allies arrived at Avenue Foch, they freed, among others, American servicemen and -women who had been working for the OSS, the forerunner of our CIA. Fifty years later, follow-up interviews with American OSS veterans who had been tortured by the Gestapo showed that 3 of 12 still suffered from well-defined symptoms of PTSD.[23]

When larger populations are studied, the percentage of those permanently scarred by PTSD, whether acquired in Nazi camps, in Vietnam, Afghanistan, or Iraq, varies widely (from about 15 percent after service in Afghanistan to over 40% in survivors of the camps). The incidence has been related to the severity of trauma, to population demographics, to social constraints—any number of factors. However, nowadays any discussion of the basic biology of post-traumatic stress disorder begins with the work of Hans Selye.[24, 25]

FROM PATHOLOGY TO PUBLICITY

Hans Selye, who introduced the world to stress and to the General Adaptation Syndrome, pursued two careers: experimental pathology and public relations. He was a success at both. As we learn from his autobiography, a genre not generally chosen by the diffident, he was born in Vienna to a respected imperial army surgeon and spent his formative years in an Austro-Hungarian cocoon of comfort, learning, and conceit.[26] He received his medical degree in 1931 from the German University in Prague—"I was the best student in my class," he confides—and earned his doctorate in organic chemistry two years later. He moved on to do post-doctoral work at Johns Hopkins and then migrated to Montreal, where endocrinology was in its heyday. After the first of his three marriages, Selye pursued a productive career in experimental pathology and endocrinology, beginning at McGill and then at the University of Montreal, where his Institute of Experimental Medicine became a major center of experimental science.

In 1936, while working on ovarian hormones at McGill, he injected impure extracts into mice and came up with the finding that made his name. He found that experimental animals responded in a stereotypic, reproducible fashion to widely diverse insults—cold, hunger, physical trauma, noxious chemicals. In rodents, either endogenous or exogenous adrenal steroids could reverse these acute disturbances.

> *We consider the first stage to be the expression of a general alarm of the organism when suddenly confronted with a critical situation and therefore term it the "general alarm reaction." Since the syndrome as a whole seems to represent a generalized effort of the organism to adapt itself to new conditions, it might be termed the "general adaptation syndrome." It might be compared to other general defense actions such as inflammation or the formation of immune bodies.*[27]

Selye went on to discover that some steroids were potent anesthetics, work that opened up the fertile area of neurosteroid research.[28] We also owe to Selye the distinction we draw today between "mineralocorticoids," like DOC, and "glucocorticoids," like cortisone.[29] His studies of calciphylaxis, a syndrome of organ-specific calcification produced by an excess of parathyroid hormone or vitamin D, not only identified a subset of human disease but provided a striking image of experimental pathology at work. The picture of one of Selye's little rodents emerging from its calcium-riddled carapace of skin made the cover of *Science*, and received major notice in the press.[30]

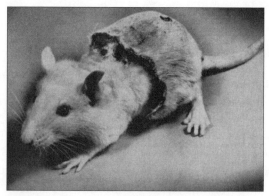

Calciphylaxis: Cutaneous molt induced by calciphylaxis in the rat.
(Selye, H., Gentile, G., Prioreschi, P., *Science*, 1961, 134:1876–1877.)

By 1946, Selye had formulated his general concept of stress and its effects on the organism: "The General Adaptation Syndrome and the Diseases of Adaptation."[31] Recognized worldwide as the Selye syndrome, its exploitation by Selye in the popular media turned an experimental pathologist into a celebrity scientist.

Nowhere in this paper, nor in his many writings, does Selye mention that stress had been introduced into experimental biology by Harvard's Walter B. Cannon in "Stresses and Strains of Homeostasis,"[32] a summary that preceded Selye's syndrome of nocuous agents by a year. Indeed, Cannon, in 1924,[33] had already implicated adrenal secretions as mediators of stress:

> *Evidence points to the sympatho-adrenal system as the chief agency in resisting alterations of our internal environment. For when that system is not functioning, the same stresses—cold, lack of oxygen, low blood sugar, loss of blood— which had no considerable influence on normal animals, become ominous for continued existence.*[34]

Be that as it may, by mid-century Selye had ventured into psychosomatics,[35] and throughout the following decades he promoted the notion of stress as the major cause, mode of transmission, and treatment of most human ills, be they mental or physical.

These quotes from the *OED* follow Selye's word "stress" from experimental pathology to psychobabble.

STRESS. *Psychol. and Biol.* An adverse circumstance that disturbs, or is likely to disturb, the normal physiological or psychological functioning of an individual; such circumstances collectively. Also, the disturbed state that results.

- *1953 Fruton & Simmonds Gen. Biochem. xxxvii. 843* Similar reduction in the adrenal ascorbic acid and cholesterol is observed when normal animals are subjected to a variety of stress [*sic*] (injury, cold, heat, drugs, toxins, lack of oxygen, etc.).

- *1955 H. Basowitz et al. Anxiety & Stress i. 7* Anxiety has been defined in terms of an affective response; stress is the stimulus condition likely to arouse such response.

- *1959 New Scientist 12 Nov. 927/1* Some examples of the diseases thought to result from stress are high blood pressure, peptic ulceration and coronary thrombosis.

- *1968 Passmore & Robson Compan. Med. Stud. II. xxxvi. 8/1* Parenthood itself can be a stress for the immature adult.[36]

By the time of his death, in 1982, Selye had written over 1,600 articles—that's about forty each year—and thirty-three books![37] These include published transcripts of his lectures on stress and disease to general practitioners, policy boffins, dentists, and proctologists. In addition to broadcasting his notions of stress, he also belabored his listeners with a "code" of moral behavior. Called "altruistic egotism," his code of behavior was a crude mix of Ayn Rand and Friedrich Nietzsche, with more egotism than altruism.

In a chapter of his autobiography called "Selling the Code"[38] he boasts that he was so well known in Canada in consequence of radio, television, and public speaking engagements that he resorted to traveling about in a wig and reflecting sunglasses, "Even I could not recognize myself," he reports, but immediately adds that—of course—he was instantly identified by a fellow passenger: "Oh Dr. Selye, how you have changed!"[39]

That would seem a fitting epitaph for a publicist who cleared the runway for Paris Hilton, but the experimental pathologist deserves better. When we eventually understand the biology of the stress syndromes, including PTSD, we'll have Hans Selye, the experimental pathologist, to thank.

5. Gore's Fever and Dante's *Inferno*: Chikungunya Reaches Ravenna

Canto XIX of Dante Alighieri's *Inferno*; print, after Sandro Botticelli.

So today, we dumped another 70 million tons of global-warming pollution into the thin shell of atmosphere surrounding our planet, as if it were an open sewer. And tomorrow, we will dump a slightly larger amount, with the cumulative concentrations now trapping more and more heat from the sun. As a result, the earth has a fever. And the fever is rising.   —AL GORE, Nobel Lecture, December 10, 2007[1]

Warming of the deep ocean and ocean surfaces appears to be altering the natural cycles that help stabilize climate over decades to millennia. . . . The explosive re-emergence of Chikungunya fever in 2004 is associated with intensifying weather extremes besetting Africa.   —PAUL R. EPSTEIN, Harvard Medical School, March 2007[2]

When they told us it was chikungunya, it was not a problem for Ravenna any more. But I thought: this is a big problem for Europe.
   —DR. RAFAELLA ANGELINI, Director of Public Health, Ravenna, Italy, December 7, 2007[3]

FEVER PITCH

LAST YEAR, GLOBAL WARMING STOPPED BEING A QUESTION OF WHETHER THE ICE caps were melting or polar bears were becoming extinct. The very week that Al Gore warned his Nobel audience that "the earth has a fever," two important studies showed how and why rising temperatures revamp the geography of infectious disease. Gore's inconvenient truth had hit the temperate zones.

In a *Lancet* publication, Rezza et al.[4] described the course of the first outbreak of Chikungunya fever in Europe. In August 2007 a sudden epidemic of this tropical disease claimed over 200 victims in and around Ravenna, Italy, a city lying on the same parallel as Bangor, Maine. Chikungunya fever is a bone-chilling, debilitating, sometimes fatal, viral illness transmitted by the bite of an infected mosquito.[5] Like several of its fellow arthropod-borne viruses (arboviruses), Chikungunya tends to affect the joints, sometimes chronically. Arboviruses launch mediators of inflammation similar to those found in several rheumatic diseases.[6] Indeed, the name "Chikungunya" derives from "kungunyala" in the Makonde language of East Africa, meaning "to double up, or to become deformed," because of joint and muscle involvement.[7] The Italian victims followed suit; one reported: "At one point, I simply couldn't stand up to get out of the car . . . I fell. I thought, O.K., my time is up. I'm going to die. It was really that dramatic."[8]

The Italian team that studied this outbreak identified the molecular structure of the virus and found it identical to one that caused earlier outbreaks in islands of the Indian Ocean. They blamed climate change for the spread into northern Italy of the virus's favorite vector, the Asian tiger mosquito (*Aedes albopictus*).[9] R. K. Pachauri, chairman of the Intergovernmental Panel on Climate Change (IPCC), who shared the Nobel Prize with Gore echoes that view. He warned that global warming has permitted the proliferation of weapons of mass infection and will affect "the health status of millions of people."[10]

AN INCONVENIENT MUTATION

The second study to appear, in December 2007, was from Stephen Higgs's group at the University of Texas Medical Branch in Galveston. They defined the molecular events that enabled Chikungunya virus to hitch a ride in the organs of *Aedes albopictus*.[11] Like the mosquito, the Chikungunya virus has also expanded its geographical range since it first appeared in Africa in the early 1950s.[12] In East Africa, the virus was limited to the habitat of its traditional vector, *Aedes aegypti*. But, after finding a more accommodating host in the tiger mosquito, the Chikungunya virus sparked new epidemics in the Indian Ocean islands of Comorros, Réunion, and Mauritius. A violent outbreak of Chikungunya fever hit Réunion Island in 2005–2006. Of the island's population of 770,000, 265,000 people took sick and 237 died.[13] WHO scientists soon found that these cases were caused by a strain of the virus that had undergone a specific mutation in the gene of an envelope protein.[14]

The Ravenna virus was carried in the blood of a South Indian man who was visiting relatives in Italy. Sure enough, the Italian doctors reported, he was infected by the mutant strain first identified in the Réunion Island epidemic.[15] Reasoning that this mutation in Chikungunya might influence its fitness for different vector species, Stephen Higgs and his team at Galveston compared the behavior of the mutant virus in *Aedes albopictus*, the vector in Réunion, and in its older West African vector, *Aedes aegypti*. They found that the mutant was not only happier in the organs of the tiger mosquito than in its traditional host, but that the virus became more infective. They concluded that "the observation that a single amino acid substitution can influence vector specificity provides a plausible explanation of how this mutant virus caused an epidemic in a region lacking the typical vector."[16]

THE MOSQUITO COAST

For an arbovirus like Chikungunya to enter a new zone, three elements are required: a change in the distribution and/or survival of the insect vector; a change in fitness and/or infectivity of the virus; and a human who carries the virus into an area deficient in mosquito control. Global warming, viral mutation, and air travel each played a role the Ravenna epidemic, while lax insect control gave squatting rights to the tiger mosquito.[17]

The tiger mosquito requires for its survival a mean monthly winter temperature greater than 0°C, a mean annual rainfall greater than 50 cm, and a mean summer temperature greater than 20°C.[18] As more regions of the globe have met these criteria, the mosquito's range has expanded. The tiger mosquito is more aggressive than most of the other *Aedes* species and tends to outbreed them. Larvae of Asian tiger mosquitoes entered Italy in 1990, tucked into stagnant coils of used rubber tires from China, via Albania—just across the Adriatic from Ravenna. *Aedes albopictus* has now been reported in ten Italian regions and nineteen provinces. Since 1984, the tiger mosquito has also made its home in much of the United States.[19] As U.S. imports of Japanese cars rose in the 1980s, so did the demand for their tires. As a result, larvae of *Aedes albopictus* arrived in the U.S. stowed in moist rubber tires destined for all those Camrys and Corollas.[20] We've been warned that this stealthy

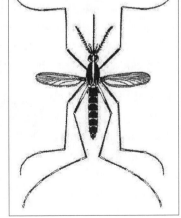

The tiger mosquito, *Aedes albopictus*.

invader—which is also the vector of dengue, West Nile virus, and eastern equine encephalitis—has set up shop in Europe and the Americas for good.[21] But why was Ravenna the first European site?

A GUELPH WAR CASUALTY

Ravenna lies in a marshy delta thirteen feet above sea level. In Roman times, Ravenna was only a few hundred yards from the Adriatic, but time and tides have moved the shores back by six miles. Effluvial reservoirs of stagnant water are guaranteed by the Corsini Canal, which connects the city to the sea, and by two small rivers, the Ronco and Montone. The site has for centuries been a safe haven for malarial mosquitoes less climatically fastidious than *Aedes albopictus*.[22]

The most famous casualty of Ravenna's malarial mosquitoes was Dante Alighieri, the immortal poet of *The Divine Comedy*. In the summer of 1321, Dante died at age fifty-six of quartan malaria contracted in the marshes around Ravenna; his tomb in the city is a major tourist site. The poet and his contemporaries knew malaria well as the dreaded quartan fever. In quartan fever, the temperature spikes every third day and is called quartan because each day of fever is counted as day one. Be that as it may, the fever is now known to be due to infection by *Plasmodium malariae*. Dante spelled out its vascular effects in the *Inferno*:

> *Like those who shake,*
> *Feeling the quartan fever coming on—*
> *Their nails already blue, so that they shiver*
> *At the mere sight of shade—such was I then . . .*[23]

Paradoxically, malaria, sometimes quartan, sometimes not, was also responsible for the growth, survival, and artistic flowering of Ravenna. Since the days of the Romans, the malarial marshes formed a natural defensive perimeter for the city's inhabitants. The site attracted settlers: "There's only one reason anyone would choose to call such a dismal place home: it was 100% defensible. To the east were treacherous shoals and lagoons, to the west were malarial swamps that effectively cut the tiny encampment off from the rest of the continent."[24] A marshy Gibraltar, as it were. Ravenna's location, facing the shores of Byzantium and straddling the northeast road to Rome, assured that waves of invaders would sweep over the land. In the millennium between AD 400 and 1500, the town fell under dominion of Visigoth and Ostrogoth invaders, of Byzantine and Lombard rulers, of Venetian doges and Roman popes.

Dante, an exile from Florence, spent his last years in Ravenna under the wing of Novello da Polenta, the region's *capo di giorno*.[25] Dante's *Inferno* begins in 1300, "In the middle of my journey of life, I came into a dark wood" (*Nel mezzo del cammin di nostra vita mi ritrovai per una selva oscura*). Alas, he never made it to the end of the journey, which at the time would have been seventy years, at best. From 1302 to the day of his death Dante was in exile from Florence, banished because of bitter factional strife that nowadays would be familiar to any observer of Middle Eastern politics—or a fan of *The Sopranos*. Medieval Tuscany was permanently divided between two rival factions, the Guelphs and the Ghibellines, and in turn, the Florentine Guelphs were further split by squabbles between white Guelphs and black Guelphs. Conflict between the two Guelph parties turned to violence, and when the black faction seized power, Dante's fate was sealed. As a sometime official under the white Guelphs, he was condemned to be burned alive if ever he returned to Florence.[26]

Diplomacy and the mosquitoes did Dante in. In 1321, he was sent by Novello da Polenta to resolve a dispute between Ravenna and Venice over some salt flats. It was a mission of little consequence and less success. One historian notes, "On his way back to Ravenna by land, for the Venetians refused him the sea passage, he caught a fever in the marshes and returned to Ravenna only to die."[27] Dante had just finished the last verses of the *Paradiso*, his apotheosis for the *The Divine Comedy*.

BEWARE THE *AEDES* OF MARSH

Mosquitoes bearing quartan fever may have been the norm in 1321, but why were tiger mosquitoes swarming around Ravenna in 2007? For reasons that remain unclear, mosquito control in the Ravenna region had lapsed back to medieval days. After World War II, energetic sanitary measures, the elimination of stagnant breeding sites, and discrete use of pesticides had driven malaria from the shores of Europe. The most effective pesticidal method was the application of 2 g/m^2 of DDT to indoor surfaces once every 6 months. A review notes: "The advent of DDT revolutionized malaria control. It enabled cheap, safe, effective treatments to be targeted at the site where most infections occur—in the home. . . . The campaign was based on a careful application of scientific principles, meticulous planning, efficient administration, generous financing, and continuous emphasis on evaluation."[28] The continent of Europe was finally declared free of endemic malaria in 1975.

A generation later, the mosquitoes came back. At the height of the Chikungunya epidemic in 2007, a Ravenna road worker complained, "In the last three or four years, you couldn't live on these streets because the mosquitoes were so bad."[29] It's not unlikely that the vigorous anti-DDT crusades of Italy's Green Party played a role. So influential has the movement become that the Greens have at times been part of the governing coalition.[30]

Whatever the error was that permitted mass resurgence of mosquitoes in Ravenna, once Chikungunya struck, immediate countermeasures were taken. Italian public health authorities sprang into action, aided by the European branch of the WHO and molecular biologists in Rome. Even before the exact pathogen was diagnosed, the public was warned to take precautions against mosquito bites, to expunge their breeding sites, and to use hotlines for tracking new cases. Pesticides were unleashed; streets, parks, and public gardens were sprayed with permethrin, while larvicides such as diflubenzoran or *Bacillus thuringiensis israelensis* were added for door-to-door disinfection.[31] So effective were these rigorous methods of mosquito control, that a public-health official could say, "By the time we got back the name and surname of the virus, our outbreak was over."[32]

Over for now, one might say. The summer epidemic may be over, but tepid tires are still being shipped around the globe and the temperate zones are fertile breeding grounds for the tiger mosquito. Any one of this sturdy brood sits ready to spread the bad news of Chikungunya fever.[33] There's a lesson for all of us to be learned from the epidemic in Ravenna, as Roberto Rosellini of the WHO told the *New York Times*: "Climate change opens the door to diseases that didn't exist here previously. This is a real issue. Now, today. It is not something a crazy environmentalist is warning about."[34] Gore was right. The effects of earth's fever are no longer limited to the Arctic and Antarctic regions. Global warming is no longer just a bipolar disorder.

6. Giving Things Their Proper Names: Carl Linnaeus and W. H. Auden

Carl Linnaeus.
Engraving by Ehrensvärd, 1740

W. H. Auden with Erika Mann, 1935.
Photograph by A. D. Bangham

Nomina si pereunt, perit et cognitio rerum
(Without names, our knowledge of things would perish)
　　　　　　　　　　　—CARL LINNAEUS, *Wästgöta-Resa, 1747*[1]

Proper names are poetry in the raw. Like all poetry they are untranslatable.
　　　　　　　　　　　—W. H. AUDEN, *A Certain World*[2]

IN 2007, WE CELEBRATED THE THREE HUNDREDTH ANNIVERSARY OF THE birth of Carl (Carolus) Linnaeus (1707–1778) whose binomial system has made it unlikely that the names of living things will perish. We also celebrated the hundredth anniversary of the birth of Wystan Hugh Auden (1907–1973), who gave his time the proper name: "The Age of Anxiety."[3]

An avatar of the eighteenth-century Enlightenment, Linnaeus set the stage for Darwin by recognizing similarities between man and ape; he named our species *Homo sapiens*. In an age when the word was unspoken, Linnaeus recognized that even plants had sex. He put Sweden on the map of natural science and changed forever the way we put names to living things.

An avatar of the Enlightened Left in the twentieth century, Auden brought English poetry into the modern world: he set planes and trains and automobiles to verse. At a time when "In the nightmare of the dark/ All the dogs of Europe bark," while Yeats was chasing the Celtic occult, and T. S. Eliot worried over "Murder in the Cathedral," Auden addressed Murder in Madrid. And after Guernica and the Hitler war, he retained high hope for a new world of reason, paying our profession a compliment we barely deserve:

> *The true men of action in our time, those who transform the world, are not the politicians and the statesmen, but the scientists. When I find myself in the company of scientists, I feel like a shabby curate who has strayed by mistake into a drawing room full of dukes.*[4]

LINNAEUS, KING OF FLOWERS

The images of Linnaeus and Auden (above) show them in the process of forging the temper of their time. Both were prodigies: Linnaeus proclaimed not immodestly, "Before the age of twenty-three, I had thought out everything."[5] He may have been right: by 1740, "*Carolus Linnaeus, Med. Doc.*" had become the founder and first president of the Royal Swedish Academy of Science; he was also practicing medicine in Stockholm. After seven years of medical study and botanical research in Lund and Uppsala, Linnaeus earned his M.D. from the University of Harderwijk, a Dutch diploma mill that is no longer extant. It was the same year he gained international attention for his *Systema Natura* (1735). Linnaeus's first effort at botanical nomenclature, at age twenty-two, had been *Præludia sponsaliorum plantarum* (*A Prelude to the Wedding of Plants*), based in good part on his 2,300-mile collecting trip through Lapland to the Arctic Ocean. But the *Systema Natura* went further into the maze of gender. He divided plants into classes by the number of "male genitals," the stamens, and then into orders by their pistils, the "female genitals": the supporting structure, the calyx, became the "nuptial bed." His sexual taxonomy may have gone a tad overboard, with some structures compared to labia minora and majora, and an entire class of flowers was named Clitoria.[6]

Whatever its sexual overtone, Linnaean botany was not only proved correct but heuristic, based as it was on specimens he himself collected worldwide. In Sweden, Linnaeus had gathered plants from Lapland to the Baltic; in Leiden, he consulted Boerhaave and gained access to Clifford's collections of the Dutch East India Company; in Amsterdam, he drew Seba's marine flora; at Oxford he scoured Sherard's global Botanical Garden; in Paris, he picked through the Jardin des Plantes with Jussieu. His botanical skills earned him a corresponding membership in the French Academy.

In 1741, a year after Ehrensvärd engraved his portrait, Linnaeus was appointed Professor of Medicine in Uppsala, and for the next three decades he and his students worked out the problem of assigning finite names to the infinite objects of nature. Key to the enterprise was a system of names. In *Philosophia Botanica* (1750), Linnaeus announced binomial taxonomy— *"A plant is completely named, if it is provided with a generic name and a specific one"* —adding that the characteristics of a species are dictated by its genus.[7] Additions and corrections to the Linnaean *Systema Natura* followed apace: the first edition of 1735 was only twelve pages long, but by the twelfth edition of 1768 it had grown to 2,300 pages and encompassed some 15,000 species. Nor was medicine given short shrift: Linnaeus first classified fevers (and hinted at contagion) in *Exanthemata Viva* (1757). Later, he extended his system to all of human disease in the *Genera Morborum* (1763). But botanical science and its application remained his first love. Dubbed the "King of Flowers" in Sweden,[8] Linnaeus made certain that if the treatment of a disease was herbal, the name of the herb was binomial.

LINNAEUS AND THE DARWINS

Linnaeus and binomial botany spread rapidly across Europe. In England, his first translator was Erasmus Darwin, FRS, Charles Darwin's grandfather. A beacon of the English Enlightenment, Erasmus Darwin was a prodigious physician, naturalist, and anti-slavery advocate. He was also a man of mirth and spirit who appreciated Linnaeus's erotic description of plants. Linnaeus had compared flowers with nine stamens and one pistil to "nine men in the same bride's chamber, with one woman."[9] Following suit, Erasmus used Linnaeus's polyandrous imagery for his own purposes in *The Botanic Garden* (1791):

> *Sweet blooms GENISTA in the myrtle shade,*
> *And ten fond brothers woo the haughty maid.*
> *Two knights before the fragrant altar bend,*
> *Adored MELISSA . . .*[10]

Charles Darwin was familiar with Linnaeus not only from his grandfather's couplets but also his own botanical studies. Half a century after *The Botanic Garden*, Charles Darwin paid homage to the King of Flowers in *On the Origin of Species* (1859):

> *Expressions such as that famous one by Linnaeus . . . that the characters do not make the genus, but that the genus gives the characters, seem to imply that some deeper bond is included in our classifications than mere resemblance. I believe that this is the case, and that community of descent . . . is the bond.*[11]

Community of human descent followed directly from Linnaean nomenclature. Linnaeus decreed that man is an Animal (kingdom), a Mammal (class), a Primate (order), and it follows, therefore, that *Homo sapiens* (genus and species) is subtended by the order of Primates. A bond, deeper than mere resemblance, would dictate that *Homo sapiens* shares Darwinian "community of descent" with apes.

Recent scholarship has shown that the decision to classify man with apes was taken jointly by Linnaeus and a fellow Uppsalian naturalist, Peter Artedi, when both worked in Amsterdam in 1735. Ironically, Artedi (now known as the King of Ichthyology) was drowned in an Amsterdam canal in the early hours of September 28, 1735, after an evening's bout of drinking with his patron, Albertus Seba. Linnaeus published Artedi's works, manuscripts, and biography, fulfilling a pact that, if one should die, then "the other would regard it as a sacred duty to give to the world what observations might be left."[12]

Erasmus Darwin picked up on the Artedi/Linnaeus notion of human origins in his *Zoönomia* and *The Temple of Nature*, only to be mocked by Coleridge for suggesting that man had "descended from some lucky species of Ape or Baboon."[13] It seems fitting, therefore, that Dr. Darwin's grandson put evolution into play once and for all. The Darwin-Wallace paper (1858) that first proposed the theory of evolution through natural selection was first read in London before members of the Linnean Society.[14]

PROPER NAMES: AUDEN AND MANN

The photograph of W. H. Auden (above) shows the poet with Erika Mann on June 15, 1935, shortly after their marriage in the registry office of Ledbury, a rural town by the Malvern hills. It was the office nearest the Down School, a prep school where Auden was teaching his last term. The snapshot was taken by Alec Bangham (better known for liposomes) and documents a change of proper name, as does an entry in the Ledbury register:

The photo and registry not only illustrate Auden's belief that proper names are poetry in the raw. They are also a document of the terrible thirties when

> *The night was full of wrong,*
> *Earthquakes and executions. Soon he would be dead,*
> *And still all over Europe stood the horrible nurses*
> *Itching to boil their children. Only his verses*
> *Perhaps could stop them . . .* [15]

Each of the names on the register reminds us of the nights full of wrong, of political earthquakes and executions. We read that Erika Julia Hedwig Gründgens, formerly Mann (occupation not stated), was married to Wystan Hugh Auden (School Master) on June 15, 1935. The registry also lists the occupation of their fathers: George Augustus Auden (Medical Practitioner) and Thomas Mann (Professional Writer), proper names indeed! Auden's father, a pioneer of public health, was an amateur of Norse myth and had named his son Wystan accordingly. The bride's father, author of *The Magic Mountain, Buddenbrooks*, etc., had won the Nobel Prize in Literature in 1929.

Marriage to Auden, a British subject, conferred a new nationality and proper name on Erika Mann, a stateless refugee from Hitler's Germany. She had arrived at the Malvern Hotel just a day before the wedding from Switzerland, where she was living in exile with her father. In Germany, she had been a film actress (*Mädchen in Uniform*), an international motorcar rally driver (Ford Motor Co.), and the emcee of a satirical anti-Nazi, gay/lesbian cabaret, the *Pfeffermühle* (à la Joel Grey in the musical *Cabaret*). Mann and her cabaret were forced to flee Germany in consequence both of her politics and sexual orientation. In April 1935, she proposed a *pro forma* marriage to Christopher Isherwood in Amsterdam as a means of gaining British citizenship. Isherwood had written *Goodbye to Berlin*, on which the musical *Cabaret* was later based, and bits of Erika Mann's career float through those Berlin stories. Isherwood declined Mann's request, pleading prior domestic arrangements. He nonetheless put Mann in touch with Auden, his school chum, sometime lover and co-author (*The Dog Beneath the Skin,* among others). Auden cabled back a one-word response to Erika's appeal: DELIGHTED![16]

Arrangements were made for a registry-office marriage near the Down school at the end of summer term. The photo and registry entries confirm the event. Directly after the wedding lunch at a local pub, Auden went back to teach classes and on the next day Erika Mann returned to Switzerland with a British passport. She telegraphed Auden almost immediately: "MEINE LIEBE, DEINE LIEBE, ALLE MENSCHEN SIND GLEICH" (my love, your love, all men are equal).[17] It's a banner worth raising against homophobes today.

Term over, Auden left the Down school forever and embarked on a soon to be well-documented voyage to and around Iceland with Louis MacNeice.[18] That trip to the mythic North of his father's dreams confirmed Auden's notion that, somewhere in this world, on some island, *Alle Menschen Sind Gleich*:

> *Fortunate Island,*
> *Where all men are equal*
> *But not vulgar—not yet.*[19]

The marriage of W. H. Auden, arguably the best English poet of the twentieth century, to Erika Mann, daughter of Thomas Mann, clearly the best German novelist of the century, is the stuff of history, raw. But a darker tone is struck by another name on the register, Gustav Gründgens, Erika Mann's first husband. Gründgens, an actor/director, was a willing player in the Nazi game of rank and honor. Greatly favored by both Göbbels and Göring, he became celebrated for his portrayal of Faust, or *Mephisto*, a role he repeated with plaudits to Nazi-packed houses in Munich and Berlin. He rose to direct Berlin's official Staatstheater during the war and produced popular musical fare for the Nazi state: one imagines a chorus of brown-shirts yodeling, *"The hills are alive with the sound of Hitler."* Always the survivor, Gründgens emerged from the fall of the Reich in a Faustian transformation, a rehabilitated, honored German hero of *Kunst*.[20] Erika took another path. She became a journalist, wrote several books with brother Klaus exposing prewar fascism, reported on the Spanish Civil War, and then wrote courtroom dispatches from the Nuremberg trials.

While Gründgens was vamping the Munich stage, others had been busy in the Munich suburb of Dachau. Erika Mann reported from the Nuremberg trials that Professors Pfannensteil of Marburg, Jarisch of Innsbruck, and Linger of Munich froze scores of inmates to death and reported detailed autopsies to "proper" scientific congresses. In Dachau, also, Professor Beiglbock of Berlin forced Poles and Jews to drink gallons of seawater: descriptions of the victims' hallucinations and heart failures were neatly recorded in what passed for scientific manuscripts. At the Natzweiler camp, Professor Dr. Eugen Haagen—formerly of the Rockefeller Institute—worked to transmit viral hepatitis from prisoner to prisoner and managed successfully to kill several hundreds with experimental typhus.[21] Pfannensteil, Jarisch, Linger, Beiglbock, and Haagen—the proper names recorded by the Nuremberg tribunals are raw poetry, indeed: *Fleurs du mal.*

AUDEN AND THE LIMITS OF SCIENCE

The Mann/Auden marriage was a noble gesture on behalf of an exile in an era when:

> *Exiled Thucydides knew*
> *All that a speech can say*
> *About Democracy,*
> *And what dictators do,*
> *The elderly rubbish they talk*
> *To an apathetic grave;*
> *Analysed all in his book,*
> *The enlightenment driven away . . .*[22]

Auden and his fellow anti-Fascists of the thirties were convinced that the journals of science contained clues to the equality of man. Auden believed that the laws of physics govern servant and master alike, and that it was the job of the poet to instruct both in the language of their common history. "Without science, we should have no notion of equality: without art no notion of liberty."[23]

Auden himself was persuaded that science, like poetry, is a "gratuitous, not a utile, act, something one does not because one must, but because it is fun." Oliver Sacks, who was an intimate, explained, "He had the analytic brilliance and vigour of a physical scientist; he had an intuitive penetrating, almost clairvoyant sense of what was going on in people, physically and spiritually, what was amiss and what was aright."[24] His oldest friends, Isherwood, Cyril Connolly, have called him a schoolboy scientist at heart; Stephen Spender acclaimed him as the diagnostician of our fears. Auden was ashamed by the extent to which the children of art and science enlisted in the service of injustice and moral squalor. Commissioned as a major at the close of the Second World War, he visited Dachau and Natzweiler, where the methods of science were mocked on behalf of

> *The grand apocalyptic dream*
> *In which the persecutors scream*
> *As on the evil Aryan lives*
> *Descends the night of the long knives*[22]

Examples of scientific disgrace were paralleled in the realm of the arts not only by the Gründgens of the stage, but also the complicities of Heidegger, the Wagnerians of Bayreuth and Oberammergau, and the films of Leni Riefenstahl. Auden was persuaded that our best chance was to establish limits to any collaboration between intellect and authority. He spelled out those hopes in his "Ode to Terminus," the Roman God of Limits:

In this world our colossal immodesty
has plundered and poisoned it is possible
*   You still might save us, who by now have*
*   learned this: that scientists, to be lucky,*
must remind us to take all they say as a
tall story . . .[25]

There's another strain here: a restatement of "without science, no equality." Auden is speaking to us from the experience of a generation that had relied on experimental science and its diversities as a shield against the biological hierarchies of Fascism. How could the brightest of Europe been deluded into surrendering equality for the grand apocalyptic dreams of one ideology or another? In "Ode to Terminus" he appealed to us to give things their proper names:

This, whatever micro-
*   biology may think, is the world we*

really live in and that saves our sanity,
who know all too well how the most erudite
*   mind behaves in the dark without a*
*   surround it is called on to interpret,*

how, discarding rhythm, punctuation, metaphor,
it sinks into a drivelling monologue,
*   too literal to see a joke or*
*   distinguish a penis from a pencil.*

One is sure that Linnaeus would have appreciated Auden's taxonomy: the last two objects are members, literally, of two different kingdoms.

7. Spinal Irritation and Fibromyalgia: Lincoln's Surgeon General and the Three Graces

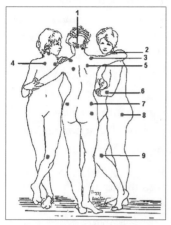

The 18 "Pressure Points" of Fibromyalgia [1]

William A. Hammond (1828–1900).
Surgeon General, U.S. Army, 1862–64 [2]

Spinal Irritation is characterized by multiple tender spots distributed over the female body, probably caused by sexual excess. . . . A couple of leeches to the inside of the nostrils are remarkably efficacious [and as for] counter-irritants, such as blisters, croton oil, tartarized antimony, and the actual cautery, cases every now and then appear in which they seem to be of service. . . . I suppose the most generally advantageous agent of the kind is the actual cautery very lightly applied to the nuchae. —WILLIAM A. HAMMOND, *Spinal Irritation*, 1886 [3]

1. History of widespread pain...
2. Pain in 11 of 18 tender point sites on digital palpation...
Digital palpation should be performed with an approximate force of 4 kg. For a tender point to be considered "positive" the subject must state that the palpation was painful. Tender is not to be considered "painful."
— "American College of Rheumatology 1990
Criteria for the Classification of Fibromyalgia" [4]

TENDER POINTS

FOR ALMOST TWO CENTURIES, DOCTORS (TRADITIONALLY MALE) HAVE BEEN responding to unexplained aches and pains of their patients (traditionally female) by tweaking various tender or painful points on their bodies. Once a particular constellation of points is elicited, a diagnosis is made and treatment applied. The women have been in real pain, current treatments have sometimes worked, but their end results have varied little over the years. Early in the nineteenth century, based on their recent discovery of the reflex arc, doctors named the syndrome of unexplained aches and pains "Spinal Irritation."[5] As described by Dr. Hammond, it was managed by applications of leeches or by hot iron cauteries applied to the back of the neck (the nuchae).

Nowadays, of course, we are much wiser. Reflexology has been left to the medically untutored and we attribute the slings and arrows of daily pain—now called fibromyalgia—to a "central hypersensitivity to nociception."[6] Experts in the field suggest that "persistent or intense nociception can lead to transcriptional and translational changes in the spinal cord and brain resulting in central sensitization and pain. This mechanism represents a hallmark of fibromyalgia and many other chronic pain syndromes, including irritable bowel syndrome, temporomandibular disorder, migraine, etc."[7]

To some with fibromyalgia, the perception of pain is so intense, its duration so debilitating, that they have been driven to extremes. A dozen years ago, Judith Curren, a forty-two-year-old nurse from Pembroke, Massachusetts, traveled to Michigan with her husband, a psychiatrist, to have her suicide "assisted" by Dr. Jack Kevorkian. Her death, according to the *New York Times*, ended "an extreme state of suffering" due to "fibromyalgia, a painful nerve and muscle disease, chronic fatigue syndrome and other opportunistic ailments."[8] The medical examiner at Dr. Kevorkian's hearing found that "there was no indication that the patient had a medical disease." The finding is in keeping with those of other medical and social scientists that there is no obvious somatic basis for the diagnosis of fibromyalgia.[9] That view has been summarized by George Ehrlich:

> The symptoms exist, some of the epiphenomena may represent contributions or reactions to the pain, and overlying everything is a sociological construct that accounts for the diagnosis and reporting in some, mainly urban, cultures, mainly in advanced and industrialized nations, and a dearth of diagnosis and reporting in developing countries and rural areas.[10]

Skeptics point out that electron microscopy of "tender points" has shown neither inflammation of connective tissue (fibro) nor of muscle (myo).[11] But whether or not fibromyalgia is a sociological construct or truly a "medical" disease, the pain (algia) that can drive a patient to suicide is real.

Since the days of spinal irritation, many of the molecular pathways and neurotransmitters of pain have been identified as targets for effective drugs.[12] In consequence, fibromyalgia is now treated with agents directed at these targets in brain and spinal cord. Recently, several such drugs have undergone double-blind, placebo-controlled studies in academic medical centers and passed the scrutiny of peer review. These include duloxetine, a serotonin and norepinephrine reuptake inhibitor;[13] tropisetron, a serotonin receptor blocker;[14] and the newly FDA-approved pregabalin, which targets the $\alpha_2\delta$ subunit of a voltage-dependent calcium channel.[15]

A puzzle arises. It seems that even homeopathic medication has survived the rigors of a double-blind, placebo-controlled study. In 1989, the *British Medical Journal* published results of a successful trial of homeopathic medicine for fibromyalgia. Supervised by an eminent English rheumatologist, the patients were given an infinitesimal dilution of *rhus toxidendron* from poison oak. The tincture had been diluted 1:99 in ethanol and "then vigorously shaken." The process was repeated six times to give a dilution of 10^{12} of the original tincture and further admixed 2 percent v/w in lactose pills. Thus by the time this "medication" had been distributed in the body no single molecule of the original material could have reached brain or spinal cord.

But a puzzle remains. Homeopathy for fibromyalgia was neither better nor worse than treatment with pharmacologically active agents. In the latest study of an FDA-approved agent, 29 percent of a treated group had their pain halved (on a scale of 1–10) versus 13 percent reduction of pain in the control group: drug had beaten placebo.[16] But the British homeopathic study reported that 53% of patients given infinite dilutions of poison oak had relief of pain and sleep disturbance versus 27 percent in the control group. Placebo had beaten out placebo.[17]

TENDER BUTTONS

The British homeopaths noted that the improvement in tenderness "which is the best discriminator" of the disease was particularly impressive.[18] Their report appeared almost simultaneously with the ACR criteria, which required doctors to tweak their patients' buttocks and nuchae with pressure sufficient to produce pain: "'Tender' is not to be considered 'painful.'"[19] In advocating eighteen bouts of hard poking at women, they were following an example set by nasty (or Nazi) predecessors in the pain game . . . as in *"Vee haff vays of making you tawk!"*

Perhaps the earliest directions for eliciting the tender points of "spinal irritation" were given by Walter Johnson in his 1849 *Essay on the Diseases of Young Women.*

The examiner stands behind the patient, and, commencing just below the neck, makes firm pressure with his knuckles successively on each projecting ridge [of] the

spinal column. Less usually he tries the effect of scalding the patient by a sponge
dipped in hot water. In the course of his investigations it frequently happens
that as soon as he presses or scalds one particular ridge or vertebra, he perceives
his patient wince or give some evidence of pain. "Aha!" says the physician,
"there it is." [20]

It is difficult to resist the Freudian connotations of this sort of examination, nowadays often conducted with the aid of a "dolorimeter," a hand-held device that measures the pressure at which tenderness or pain is provoked.[21] Some of the fibromyalgia doctors have broadened their purview by means of a similar device, a "vulvodolorimeter," which is used to measure pain in the female genitals.[22] One is reminded of Gertrude Stein's *Tender Buttons,* a volume in which "slyly erotic references" to "pain soup" celebrate her domestic arrangements with Alice B. Toklas.[23]

THE THREE GRACES

The astonishing image of the Three Graces that illustrates the ACR criteria (see above) reflects neither the somatotype of Gertrude Stein nor of most sufferers from fibromyalgia. It's been noted that patients afflicted with fibromyalgia, as were patients with "spinal irritation" in years past, are significantly more corpulent than their sisters. Indeed, both the number of tender points and the extent of pain in fibromyalgia tend to vary with the body mass index.[24] Poor Judith Curren, who died for her disease, was doubly afficted. "I think this one was a tragedy," the medical examiner told reporters. "I have no doubt she was tired. When someone is 260 pounds, you easily get tired. When someone is depressed, you get tired." Mrs. Curren was five feet, one inch tall.[25]

An argument in favor of the social construct theory is as handy as the Google image bank of the Three Graces. Depictions of those paragons of female grace and beauty, of Aglaia (the radiant), Euphrosyne (the joyful), and Thalia (the flowering), vary as widely with time and fashion as society itself. The ACR trio is a very trimmed-down, 1990 version of a far plumper set of Graces by Henri Alexandre Georges Regnault, now in the Louvre.[26] The Regnault ladies advertise the glorious results of wining and dining in mid-nineteenth-century France. Far slimmer Graces were carved in stone earlier in the century by Thorvaldsen and Canova; the neoclassic period favored cleaner lines and smaller poitrines.[27] The neoclassic Graces are eclipsed in BMI by the gusty Flemish amazons of Peter Paul Rubens;[28] his age favored fuller bodies and flacons of brew. As ever, the Renaissance combined best the flowering of beauty with the joy of the radiant: the Graces of

Raphael and Botticelli are forever immortal.[29] The squalid present day of twenty-first-century art and dolorimetry is well represented by the grotesques of Philip Pearlstein and the hermaphrodites of Joel-Peter Witkin.[30] We might conclude that society dictates the form and function not only of real, but also ideal, women.

LINCOLN'S SURGEON GENERAL

Edward Shorter addressed the construction of disease in the context of spinal irritation, focusing on the days of Regnault and William Hammond:

> *Reflex theory . . . claimed a scientific basis in the pseudoscientific doctrine of "irritation" and in the genuinely scientific notion of reflex arcs in the spine. Yet the popularity of reflex theory among physicians was probably a result of their pejorative beliefs about women—beliefs that male physicians held in common with other males of the nineteenth century.[31]*

The spinal irritation and counterirritation reached its zenith in the 1886 book *Spinal Irritation*, by Dr. William A. Hammond of New York, who routinely found "multiple tender spots" in women suffering from spinal irritation and ascribed some of the cases to sexual excess or, so he was convinced, masturbation. In keeping with his contemporaries, he advocated treatment with "counterirritants" such as dry heat, scalding water, or croton oil extracts.

William Alexander Hammond had been for a short time Lincoln's Surgeon General of the United States. He attained the distinction, unique for that rank, of being court-martialed during the Civil War after acrimonious squabbles with Secretary of War Stanton and bureaucrats of the career Medical Service.[32] His chief problem, aside from a personal lapse in petty finance, seems to have been that he was too closely affiliated with the reformers of the Sanitary Commission. But Hammond was able to recover from political infamy; eventually his reputation was cleansed and his rank restored by the Senate. He went on to become one of the founders not only of the American Neurological Association, but also of the NYU Postgraduate Medical School, in the library of which his books are now quietly disintegrating.

Hammond's magisterial *A Treatise on Diseases of the Nervous System* (1871), the first American textbook of neurology, is an extensive tome, published a scant seven years after his fall from official grace. The treatise is filled with outmoded rituals and jaw-breaking syndromes. Scattered among this dross are neatly described case histories and new observations, but the volume is tough slogging. Much like the folk who today play with "dolorimeters" or "vulvodolorimeters," Hammond had his own, patented machine for neurological diagnosis, the "dynamograph."[33]

On April 25, 1870, at the court of General Sessions in New York City, Hammond testified on behalf of Daniel McFarland, a patient of his who had shot the famous journalist Albert Richardson. Hammond told the jury that he had, "devoted the last five years of his professional life exclusively to the study of the mind" and had diagnosed McFarland's cerebral hyperemia by means of his "dynamograph" machine. Hammond assured the jury that the machine's test of the accused's motor responses provided "full and decided evidence" that McFarland could not control his will."[34]

Among his patients with "spinal irritation" was poor Miss A. W., whose diagnosis was tough to unravel. One of her symptoms was the habit of swallowing pins by the dozen to stop her pains. Hammond discusses the puzzling diagnosis and, in charged phrases, describes a neurologist arriving at the true diagnosis:

> . . . he is baffled, but then instead of [giving up], he will begin to punch or hammer the vertebrae, as he before pressed them. In this way it very rarely happens but that he at last succeeds in finding some sensitive spot, which he can assume to be the seat of the disease. He now feels it a clear duty to apply leeches to the culprit vertebra, or mercurial inunction, or a blister, or an issue or seton [a few silk threads inserted into a surgical incision in the skin, to excite pus], and strictly enjoins perfect quiet and the recumbent position.[35]

Happily, and perhaps in consequence, Miss A. W. later extruded these pins from her skin, her nose, and various nether orifices (surely the first documented case of pins envy).

Nevertheless, Hammond showed a degree of empathy for his mentally ill patients, regarding them in the tolerant, bemused fashion of a Victorian author displaying his fictional creatures. But woe to the patient who failed to respond to "moral means," the talking cure. She—for it was usually a female patient—was in for leeches or counterirritants. Hammond was convinced that all neurological diseases were due to either "hyper-" or "hypo-emia" and that the cautery and leeches would restore the balance between them. Since mainly women were subject to these measures, it is fair to ask whether sexism, historical folly, or medical science was at work.

The world of Hammond and the spinal irritation doctors seems so far away, viewed from our privileged decade of molecular medicine, of sono- and angiograms, of cyclo- and cephalosporins, of liposomes and liposuction. The day-to-day medicine of the 1830s to the 1930s appears to have been futile, groping, and—let's face it—quackish. All that laying on of hands, thumping of the chest, leeches, calomel, glycerine, cautery! In 1871, Oliver Wendell

Holmes gave the students at Hammond's medical school a vision of medical science which transcends the fashionable diagnoses and drugs of the day; it will remain true when dolorimetry will have gone the way of the cautery, and pressure points the way of calomel:

> *If the cinchona trees all died out . . . and the arsenic mines were exhausted, and the sulphur regions were burned up, if every drug from the vegetable, animal, and mineral kingdom were to disappear from the market, a body of enlightened men, organized as a distinct profession, would be required just as much as now, and respected and trusted as now, whose province should be to guard against the causes of disease, to eliminate them if possible when still present . . . and to give those predictions of the course of disease which only experience can warrant, and which in so many cases relieve the exaggerated fears of sufferers and their friends or warn them in season of impending danger.*[36]

8. Tithonus and the Fruit Fly: New Science and Old Myths

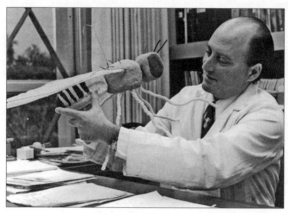

Seymour Benzer (1921–2007), with wooden model of *Drosophila* (ca. 1974)[1]

Vase: Eo pursuing Tithonus[2]

POPE BENEDICT XVI SAYS AN IMMORTALITY PILL MIGHT NOT BE SUCH A GOOD THING. (Vatican City) *The 80-year-old pontiff says it's better not to hope for biological life that can be made to last forever.*   —ASSOCIATED PRESS, March 3, 2008[3]

The Drosophila mutant methuselah (mth) *was identified from a screen for single gene mutations that extended average lifespan. Mth mutants have a 35% increase in average lifespan and increased resistance to several forms of stress.*
     — "Crystal Structure of the Ectodomain of Methuselah"[4]

I don't want to achieve immortality through my work; I want to achieve it through not dying. I don't want to live on in the hearts of my countrymen; I want to live on in my apartment.   —WOODY ALLEN[5]

METHUSELAH AND THE "SUN"

FOR REASONS CLEAR TO HIS HOLINESS BENEDICT XVI, IF NOT TO WOODY Allen, more folks bet on eternal paradise than on permanent rent control. Last March, while delivering a homily near St. Peter's Square, the Pope reflected on the limits of science. Pleading for the faithful to "drink from the fountain of life itself," i.e., spiritual immortality, he warned that extending life past its natural limits would crowd the world with old people. There would be no room on earth "for youth, for this newness of life."[6] I'm afraid, though, that his warning was a tad late; modern science has already helped break the biblical rule that "the days of our lives are three score years and ten" (Psalms 90:10). The Pope has himself beaten the odds. Thanks to advances in medical management, the octogenarian pontiff has survived two strokes and congestive heart failure.

Biological immortality might not lurk around the corner, but we're working on it. In the lab, thanks to the biological revolution, we've "immortalized" lines of human cells in a dish, cloned DNA from the dead, and extended by tenfold the natural life span of yeasts.[7] Flies, worms and rodents live up to 40 percent longer when fed a diet that has at least 30 percent fewer calories than usual,[8] while resveratrol from red wine can keep them fit.[9] Reporting on Leonard Guarente's analysis of how caloric restriction extends life, the *New York Times* enthused:

> *Actuaries, put down your gloomy mortality tables and sharpen your pencils. Heirs and legatees, contain yourselves in patience. If any such drug were to work in humans the same way that this diet of 30 percent less than normal calories works in laboratory rodents, people would start enjoying a maximum life span of 170 years, most of it in perfect health.*[10]

And sure enough, news came last summer that an immortality pill is nigh. *Science Daily* reported that "SUPER FRUIT FLY MAY LEAD TO HEALTHIER HUMANS; AGING SLOWED WITH SINGLE PROTEIN."[11] Richard Roberts of USC and Seymour Benzer's group at Caltech have made peptides that target the Methuselah (mth) gene in fruit flies. Insects with mutations in this gene have a 35 percent increase in average life span and increased resistance to several forms of stress, including heat, starvation, and damage by oxygen excess. The protein affected by this mutation is related to a gut-hormone receptor in humans that is switched on by a ligand named "Sun." By means of a novel selection technique which Roberts developed, the Pasadena group identified peptide inhibitors that bind to the mth protein and prevent binding of "Sun" to its target.[12] Since humans, like flies, have these proteins on many of our cells, this inhibitory sunscreen—so to speak—could be a forerunner to the pontiff's "immortality pill." As Woody Allen might say, we should live so long.

THE EVENING HEMISPHERE

We've actually been heading in that direction for more than a century. In the 1850s, American life expectancy was 39.5 years, and poet James Russell Lowell complained:

And yet you ask me why I'm glum,
And why my graver muse is dumb,
Ah me! I've reasons manifold
Condensed in one—I'm getting old
When life, once past its fortieth year
Wheels up its evening hemisphere . . . [13]

These days, age 40 seems closer to noon than evening. American life expectancy is now nearly 80, and our maximum life span has also increased. The bean counters of Medicare predict that by 2010 they'll have 50,000 centenarians on their payroll.[14] The trend is global: the oldest Swedes now die at age 108, eight years later than their counterparts in 1860.[15] The days of our lives are no longer three score years and ten.

A Chinese official boasts that "Chinese Life Expectancy Rises by 41 Years in One Century," from just over 30 years in 1900 to 70.1 in 2000. He attributes the new longevity of China's people to "the advancement of science and technology, especially in medical science."[16] *Pace*, Pope Benedict—the Chinese young have not been crowded out by old folks. When the pontiff urged a moratorium on immortality pills, I'd bet that he was hooked up to a sound system manufactured by one of the 300 million Chinese under 25 who now walk the earth because of one or another pill.

The longer the days of our lives, the longer the days of waning powers. There's that Catskill story about the 87-year-old man who marries an 83-year-old woman whom he met in a senior center. They take their honeymoon in a rented bungalow by the sea. The woman retires upstairs first and calls out to her husband, "Come upstairs and make love!" He replies, "I can't do both!"

NO COUNTRY FOR OLD MEN

The upside? "Bodily decrepitude is wisdom; young/We loved each other and were ignorant," insisted William Butler Yeats.[17] Clearly, the more folks hobble and creak, the more likely they are to conflate their nostalgia for youth with a pipe dream of paradise. Yeats found that vision in the Byzantine mosaics of Ravenna:

That is no country for old men. The young
In one another's arms, birds in the trees
—Those dying generations—at their song,
The salmon-falls, the mackerel-crowded seas,
Fish, flesh, or fowl, commend all summer long
Whatever is begotten, born, and dies.

Caught in that sensual music all neglect
Monuments of unageing intellect.

—W. B. YEATS, "Sailing to Byzantium," 1928[18]

Yeats was sixty-three years old when he wrote "Sailing to Byzantium" and like poets before and after, equated longevity with decrepitude. He mourned that "An aged man is but a paltry thing,/ A tattered coat upon a stick," and that his heart was "fastened to a dying animal." But there was pie in the sky of Byzantium. Yeats looked forward to being gathered "into the artifice of eternity" where in avian form he would be

. . . set upon a golden bough to sing
To lords and ladies of Byzantium
Of what is past, or passing, or to come.[19]

Yeats resigned from the Irish Senate for "reasons of health" in the year that the poem was published. Tattered and paltry, perhaps, but he stuck out the years of hobble and creak for eleven more years, until in W. H. Auden's words, "The current of his feeling failed; he became his admirers."[20] His admirers have kept Yeats very much alive: the film *No Country for Old Men* recently picked up four Oscars; Jonathan Spottiswoode's rock lyric "Sailing to Byzantium" is climbing the charts; and Philip Roth's *The Dying Animal* is in all the bookstores.[21] To what avail? While Yeats may have achieved immortality through his work—and lives on forever in the hearts of men—he's no longer living in his apartment.

FREUD, TITHONUS, AND TENNYSON

Unlike Yeats or the pontiff, Sigmund Freud was as skeptical of the afterlife as Woody Allen. At mid-life in the Vienna of 1911, Freud scoffed at spiritual immortality: "The doctrine of reward in a future life for the renunciation of earthly lusts is nothing but a mythical projection."[22] But age and illness convinced Freud that the land of pain is no country for old men. Age eighty and exiled in London, mauled by bouts of surgery and radium treatment for oral cancer, Freud turned to the kindly Fates of Greek myth:

Perhaps the gods are kind to us in making life more disagreeable as we grow older.
In the end death seems less intolerable than the manifold burdens we carry.[23]

As he wrote these lines in his study, Freud was attended by a goddess. Eo, goddess of dawn, is limned in clay on an Athenian *Lekythos* which stood by his desk. This red-figured oil jug (see above), which dates to the fourth century BC, remains on view today in London's Freud Museum, while its name,

Lekythos, remains in our vocabulary as lecithin, from the Greek for egg yolk.[24] The graceful, spread-winged goddess is shown in pursuit of a beautiful youth with a lyre in his hand. The legend is that of Eo and Tithonus, a handsome mortal beloved by Eo. The goddess was so enamored of the young prince of Athens that she asked Zeus to make Tithonus immortal. He did, but she failed to ask Zeus to prevent Tithonus from aging. As time went on—as the Fates wound down the string of life—he did indeed turn into a paltry thing: he aged, dried and shrank to the size of a cricket. And that's why crickets chirp at dawn.

Over the years the legend has been iterated in poetry, music, and the visual arts; as our life span increases, each telling is more poignant. A version pertinent to modern biology was written in 1860 by Alfred, Lord Tennyson. The poet has Tithonus pleading with Eo at dawn's early light:

> *How can my nature longer mix with thine?*
> *Coldly thy rosy shadows bathe me, cold*
> *Are all thy lights, and cold my wrinkled feet*
> *Upon thy glimmering thresholds, when the steam*
> *Floats up from those dim fields about the homes*
> *Of happy men that have the power to die, . . .*[25]

It's clear that what rickety-crickety Tithonus needed was a pill that would not only make him immortal, but keep him young enough to mix his nature (as it were) with Eo's. If only he had been around to hear the news from Caltech! Shrunken to insect stature, what would Tithonus not have given for a mutated mth gene, for that peptide inhibitor, for that pill from Pasadena!

AFTER MANY A SUMMER DIES THE SWAN

But wait. Pasadena? Tennyson? Tithonus? I seem to have heard this song before. It turns out that one of the best-known works of longevity fiction is actually set in Pasadena: Aldous Huxley's *After Many a Summer Dies the Swan* (1939). Huxley, grandson of the great Darwinist Thomas Henry Huxley, took the title from the opening lines of Tennyson's "Tithonus."

> *The woods decay, the woods decay and fall,*
> *The vapours weep their burthen to the ground,*
> *Man comes and tills the field and lies beneath,*
> *And after many a summer dies the swan.*[26]

The action of the Huxley novel takes place a "short motor-car ride" north of Los Angeles. On this as-yet-undeveloped site, a boorish tycoon has built a large mansion (half the San Simeon of Hearst, half the Huntington Library) that houses valuable ancient manuscripts. An English scholar is sent to look into a sheaf of books and manuscripts that the mogul has acquired from an English earl; these date back over 300 years. The tycoon has hired a certain Dr. Obispo [*sic*] to inject him with an extract that will prolong his life. The extract is prepared from the intestinal flora of carp—whose life expectancy far exceeds that of humans. Obispo's biological experiments are based on the notions of Elie Metchnikoff, who won the Nobel Prize in 1908 for discovering phagocytosis. Metchnikoff suggested that we age because our intestinal bacteria generate toxic, oxygen-derived metabolites of fatty alcohols; these activate phagocytes of all stripes: macrophages and microphages (read neutrophils) and "neuronophages" (read glial cells). Dr. Obispo tests the effects of feeding carp intestines to mice: "The effect on the mice had been immediate and significant. Senescence had been halted, even reversed. They were younger at eighteen months than they had ever been."[27] Just like Guarente's starved mice or Benzer's fruit flies.

The locale also rings a bell. It seems that Huxley's fictional mogul is the chief supporter of an equally fictional university, Tarzana U, about to be erected near that monumental archive of manuscripts. The founder of Tarzana, Dr. Mulge, brags to his donor that "the Athens of the twentieth century is on the point of emerging here in the Los Angeles Metropolitan area. I want Tarzana to be its Parthenon [for] Art, Philosophy, Science."[28] Tarzana seems to be the fictional stand-in for Caltech; Huxley in 1939 was living in Los Angeles and experimenting with hallucinogens that might spur the imagination. The seminal spirit of Caltech was George Ellery Hale, an astronomer and the first director of the Mount Wilson Observatory. The real Dean Hale foresaw the future development of his Parthenon in Pasadena: "No creative work, whether in engineering or in art, in literature or in science, has been the work of a man devoid of the imaginative faculty."[29]

Aldous Huxley would have been pleased that the California Institute of Technology is within a stone's throw of the Huntington Library, where from time to time is displayed a copy of the *Ellesmere Chaucer*, acquired by Henry Huntington in 1917 from the third Earl of Ellesmere, whose family had owned it for 300 years. Huxley's Dr. Obispo might have been pleased, were he to have learned that nowadays we, too, link aging to oxygen-derived free radicals. Indeed, we now have experimental evidence that many receptors like those described by Benzer et al. are involved in phagocytosis and immunity, and not only to that gene for longevity in the fly.[30] Who would have thought it?

Well, William Blake for one, in his poem "The Fly":

Am not I
A fly like thee?
Or art not thou
A man like me?

For I dance
And drink & sing:
Till some blind hand
Shall brush my wing.

If thought is life
And strength & breath,
And the want
Of thought is death;

Then am I
A happy fly,
If I live,
Or if I die.[31]

We have learned much about the life of flies and men, about Methuselah and Sun, and about other genes for "Time, Love, Memory"[32] from the wise man who put the fruit fly on the map of Pasadena forever: Seymour Benzer (October 15, 1921–November 30, 2007). He, too, has become his admirers.

9. Swiftboating "America the Beautiful": Katharine Lee Bates and a Boston Marriage

"The Republic" (1893), by Daniel Chester French.
Columbian Exposition, Chicago

Katharine Lee Bates (1859–1929).
Poet of "America the Beautiful"

O beautiful for spacious skies
For amber waves of grain,
For purple mountain majesties
Above the fruited plain!
America! America!
God shed his grace on thee
And crown thy good with brotherhood
From sea to shining sea!

—KATHARINE LEE BATES, "America the Beautiful" (1893)[1]

I strongly believe the neglected American people need . . . leadership and our
Country needs to return to America The Beautiful in every way possible.

—LINDA ARCHER, Reader comment on washingtonpost.com
"Post I.T." blog April 18, 2007[2]

AND CROWN THY GOOD WITH BROTHERHOOD

IN THE SUMMER OF 2007, AS POLITICAL TEMPERS FLARED IN EARLY SKIRMISHES from Iowa to the Carolinas, one rather nasty theme emerged. IOWA GAY MARRIAGE RULING STIRS 2008 RACE ran the headlines, and a contest started for the bipartisan laurels of bigotry.[3] And sure enough, between mug shots at weenie roasts and platitudes at county fairs, a handful of hopefuls warned the faithful that marriage between people of the same sex ranked among the major threats to our republic.

Irony rampant: the same cameras that showed us politicians of every stripe and party desporting themselves at the Iowa State Fair also featured squeaky-clean farm kids welcoming visitors to the fair with the rousing verses of "America the Beautiful." So on behalf of a good number of my fellow citizens—and of their legislators as well—I'd like to remind both present and future candidates for office of Katharine Lee Bates, who wrote the poem "America the Beautiful." It's the story of a happy Boston Marriage in the Era of the White City.

O BEAUTIFUL FOR PILGRIM FEET

Katharine Lee Bates (1859–1929) is the most famous native of Falmouth, Massachusetts; her statue decorates the library lawn, the road to the library bears her name, the bicycle path along Vineyard Sound to Woods Hole is named "The Shining Sea," and the upscale granola store is called "Amber Waves." Her poem "America the Beautiful" is usually sung to music set by Samuel A. Ward, a Son of the American Revolution. It pays homage to their New England forbears:

> O beautiful for pilgrim feet,
> Whose stern, impassioned stress
> A thoroughfare for freedom beat
> Across the wilderness!
> America! America!
> God mend thine ev'ry flaw,
> Confirm thy soul in self control
> Thy liberty in law! [4]

It's a sentiment less bellicose than that expressed in our official national anthem—"the bombs bursting in air"—and considerably sweeter than the boast of "*Deutschland Über Alles*," the pomp of "God Save the Queen," or the gore of "*Le Marseillaise*." It is also a fitting postbellum sequel to Julia Ward Howe's "The Battle Hymn of the Republic." Indeed, Julia and Samuel A. Ward were Yankee kin. [5]

Bates was inspired to write "America the Beautiful" on her first trip out West. A professor of English at Wellesley, she had been asked to teach English religious drama at a summer school in Colorado Springs and spent a happy "three weeks or so under the purple range of the Rockies." [6] To celebrate the end of the session, she and others on the faculty made an excursion to Pike's Peak, pulled to the summit by mules in prairie wagons that bore the slogan "Pike's Peak or Bust!" "It was at the summit, as I was looking out over that sea-like expanse of fertile country spreading away so far under those ample skies, that the opening lines of the hymn floated into my mind," she recalled.[7]

She left Colorado Springs with notes for the entire four stanzas and other memorabilia of her extended trip to the Rockies, but the poem did not appear until July 4, 1895 in *The Congregationalist.* After a musical setting by the once well known Silas G. Pratt attracted national attention, the popular stanzas became open game for other musical versions, and by 1923 more than sixty "original" settings had been perpetrated. The verses can be sung to many old tunes, including "Auld Lang Syne" and "The Harp That Once Through Tara's Halls." But the setting we know best nowadays was adopted by Ward from the hymn "Materna" and the words we use are those of Bates's revised version of 1913.[8]

CONFIRM THY SOUL IN SELF CONTROL

"Thy liberty in law!" could serve as a motto for Bates and her impassioned generation of pilgrim daughters. Bates was graduated from Wellesley in 1880, ten years after that stern, seminary-style college had been chartered as a place for "noble, white unselfish Christian Womanhood." But by 1882, winds of change from the West brought a new generation in the person of Alice Freeman (1855–1902). Freeman was only twenty-seven years old when she was called from the coeducational University of Michigan to become Wellesley's second president. She proceeded to transform Wellesley into a college ready for the twentieth century; she also helped to found Radcliffe and the Marine Biological Laboratory.

As president of Boston's Woman's Education Association (WEA), Alice Freeman collaborated with Elizabeth Cary Agassiz, widow of the biologist, to work out the legal arrangements whereby the "Harvard Annex for Women" became Radcliffe College. The WEA also raised $10,000 to promote the teaching and research by women in science. This gift made it possible in 1888 to purchase land near the all-male U.S. Fisheries building at Wood's Holl (as it was then) to establish the Marine Biological Laboratory. The WEA also insured that women might work at the laboratory by requiring the presence of two of its members on the board of trustees: the first two were graduates of Vassar and MIT.[9]

Among Freeman's first appointments at Wellesley were Eliza Mosher as professor of practical physiology and Katharine Coman (also from Michigan) as professor of political economy and history. In 1885, she appointed Katharine Lee Bates an instructor of English. They were soon joined by Mary Calkins, a student of William James, who established the first laboratory of experimental psychology to be headed by a woman. Like Mosher, Coman, and Calkins, Katharine Lee Bates was destined to spend her entire life at Wellesley; she became full professor in 1891 and long-term chairman of the English Department until her retirement in 1925. But her life changed forever in 1887 when she met Katharine Coman. The two Katharines lived together for more than a quarter of a century in the loving bonds of what was then called a "Boston marriage" and is now appreciated as "a devoted lesbian couple."[10] They called their Wellesley home the "Scarab," their faithful collie "Sigurd," and their automobile "Abraham" (because they were so often deep in its bosom). When parted by professional travel, they wrote passionate, almost daily, letters to each other:

> *Your love is a proof of God. How does love come, unless Love is? . . . That is a glorious sentence wherewith to close your letter. I love it and I love you and I love what shadowy hint of God comes to me.*[11]

THINE ALABASTER CITIES GLEAM

"Central Fountain of the Columbian Exposition," by Frederick W. MacMonnies (Chicago 1893).

In 1893, on that journey to Pike's Peak, the two Katharines stopped to visit the great Columbian Exposition in Chicago, Bates becoming "naturally impressed by the symbolic beauty of the White City." By that time Alice Freeman had become Alice Freeman Palmer, after marrying a Harvard philosopher, and the Palmers together supervised construction and installation of the Woman's Building, a monument fashioned in the mock alabaster of the exposition's Beaux Arts style. Featuring statues of prominent feminists such as Susan B. Anthony and Elizabeth Cady Stanton, its interior was chock -a-block with objects picked to show:

> . . . the contributions made by women to the huge workshop of which this world so largely consists, their contributions not only to the industries of the world but to its sciences and arts. Thus it is hoped in a measure to dispel the prejudices and misconceptions, to remove the vexatious restrictions and limitations which for centuries have held enthralled the sex.[12]

The Woman's Building was an anomaly among the grander structures of the Columbian Exposition, the iconography of which played to the "prejudices and misconceptions" of centuries. In statuary great and small, burly men steered the ship of state, while women were placed on pedestals, perched as guiding spirits or cast as docile handmaidens. MacMonnies' "Central Fountain of the Columbian Exposition" underscored each of these roles.[13]

A visit to the Palmers' Woman's Building was not the only reason why the two Katharines had stopped in Chicago; there were professional reasons as well. Bates had taught in Colorado Springs on equal terms with such male professors as Rolfe of Harvard and Todd of Amherst, while Coman had lectured on the economic history of Western expansion. In Chicago, Coman heard Frederick Jackson Turner deliver his famous address on "The Role of the Frontier in American History." He argued that the Western frontier had for three centuries been a metaphor for the American dream, for "manifest destiny." Now that the open frontier had closed and the United States had become one nation from sea to shining sea, other frontiers awaited. Coman's two-volume *Economic Beginnings of the Far West* was devoted to parallel themes, and her economic history of the railroads again echoed Turner's message that the age of an external frontier and the West as wilderness was over. Coman and Bates were convinced in 1893 that the spirit that had won the West would in time remove those "vexatious restrictions and limitations."[14]

GOD MEND THINE EV'RY FLAW

The America of continental expansion was no collection of white cities. 1893 was a year of significant social unrest, and the strife was by no means liberating. Grover Cleveland entered the White House for the second time, with the country in the midst of a deep economic depression. On February 23, 1893, the Philadelphia and Reading Railroad had gone bankrupt, and before the end of the year the Erie, the Northern Pacific, the Union Pacific, and the Atchison, Topeka and Santa Fe went belly up as well. Two and a half million people were unemployed—one fifth of the work force—and Henry Adams lamented that "much that had made life pleasant between 1870 and 1890 perished in the ruin."[15] Even the president admitted that "values supposed to be fixed are fast becoming conjectural, and loss and failure have invaded every branch of business." In legislation that was to provide windfall profits for J. P. Morgan and for Augustus Belmont, Cleveland brought the country back to the gold standard in the very week that Bates stood atop Pike's Peak.[16]

Meanwhile, federal and state militias were sent against workers in the Carnegie/Frick (or Homestead) steel strike, against switchmen in Buffalo and coal miners in Tennessee, and finally into the Pullman strike in Chicago. Each of these episodes of class warfare was later treated in Katharine Coman's *Industrial History of the United States* (1905).

The two Katharines were also in the vanguard of social activists. Bates and Coman were among the founders in 1887 of the College Settlements Association, a group that made it possible for young female college graduates to spend a year at community settlement houses among the poor and the immigrants—the "teeming refuse" of Europe's shores. In the course of this work Bates and Coman became closely associated with the pioneer of Chicago's Hull House—and future Nobel Peace Prize winner—Jane Addams. In 1889, Addams's lifelong companion, Ellen Gates Starr, described how the settlement-house movement might benefit not only the needy but also the philanthropist: "It is not the Christian spirit to go among these people as if you were bringing them a great boon: one gets as much as one gives [but] people are coming to the conclusion that if anything is to be done towards tearing down these walls . . . between classes that are making anarchists and strikers the order of the day, it must be done by actual contact and done voluntarily from the top."[17] A generation of social workers, public-health activists, and egalitarians spent their lives convinced of the need for that actual contact.

In 1892, Katharine Coman became chairman of the committee that opened Denison House in Boston and made it a center of labor organizing activity, to which Bates was inevitably drawn. Denison House, Hull House, and the other settlement houses were deeply committed to reform of working hours, protection of immigrants, compulsory school attendance, school health and—above all—abolition of child labor. It was toward this end that the poet Sarah Cleghorn wrote in "The Masses":

The golf links lie so near the mill,
That almost every day,
The laboring children can look out
And see the men at play.[18]

When violence broke out during the Chicago Pullman strike of 1894, and strikers burned down the remnants of the White City, Coman and Addams sided with workers against the militia; Coman went to Chicago again in 1910 to help striking seamstresses win union rights.[19]

UNDIMM'D BY HUMAN TEARS

Meanwhile tragedy had struck the couple; Coman's last work was completed as she lay dying of breast cancer. *Unemployment Insurance: A Summary of European Systems* (1915) was a meticulous survey of how other industrialized countries cared for the aged, the disabled, and the unemployed.[20] She concluded that social services in Bismarck's Germany and the Third Republic's France were far in advance of those in her own country. Posthumously published to little acclaim, Coman's book was to become a blueprint for social justice in the United States. Coman's call for old-age and disability benefits in the New World—social security—became a platform plank of the Progressive and then the Democratic party. After the New Deal, Coman's dream became the law of the land. Bates describes how she helped her friend finish this major economic study:

Through those four years beset with wasting pain,
The surgeon's knife again and yet again,
. . . So we twain
Finished your book beneath Death's very frown.
For all the hospital punctilio,
Through the drear night within your mind would grow
Those sentences my morning pen would spring to meet . . .[21]

The lines come from a volume of passionate love poems, *Yellow Clover,* written by Bates and published in 1922, seven years after Coman died. Each of these poems was devoted to Katharine Coman and in some Bates reached levels of emotional expression—perhaps even art—that eluded her in seven other volumes of verse.

Your life was of my life the warp and woof
Whereon most precious friendships, disciplines,
Passions embroider rich designs . . .

No more than memory, love's afterglow?
Our quarter century of joy, can it
Be all? The lilting hours like birds would flit
By us, who loitered in the portico
Of love's high palace . . .[22]

Bates spoke in no loud voice the love that dared not speak its name: yellow clover stood for physical love in the flower language of the two Wellesley scholars, who

Stooped for the blossoms closest to our feet
And gave them as a token
Each to each
In lieu of speech,
In lieu of words too grievous to be spoken . . .[23]

"Undimm'd by human tears" is the hopeful lyric of Bates' most successful public poem, our national anthem of social justice, a hymn to the better angels of American nature. The lyrics of "America the Beautiful" should remind the bigots of this world of a generation of women whose emotional ties and social reforms have outlasted the alabaster cities' gleam of the Columbian Exposition.

And crown thy good with brotherhood
From sea to shining sea!

10. Nothing Makes Sense in Medicine Except in the Light of Biology

The Lesson of Claude Bernard (1884), by Léon Augustin L'Hermitte

Judah Folkman (1933–2008)

Nothing in Biology Makes Sense Except in the Light of Evolution.
—THEODOSIUS DOBZHANSKY, 1973[1]

For a man of science there is no separate science of medicine or physiology, there is only a science of life. —CLAUDE BERNARD, 1865[2]

By viewing the process of angiogenesis as an "organizing principle" in biology, intriguing insights into the molecular mechanisms of seemingly unrelated phenomena might be gained . . . —JUDAH FOLKMAN, 2007[3]

MEDICINE DOES NOT END IN HOSPITALS. . .

WHEN JUDAH FOLKMAN DIED EARLIER THIS YEAR, A GENERATION MAY have passed. His was the last cohort of chief resident physicians and surgeons to make sense of medicine in the light of experimental biology. Products of the post–World War II era, when hands-on benchwork was the rule in medical schools, Folkman's generation was given further training in labs run by the Army, Navy, or the NIH. They returned to universities where scientific curiosity and laboratory experience—perhaps even discovery—were expected. It may well have been "the greatest generation" of medicine; its members presided over an era when students flocked from every corner of the globe to learn experimental medicine in America.

These days, sad to say, modern "health care" (a.k.a. medicine), with its competing demands of evidence-based practice, center-based bureaucracy, drug-based clinical trials, gadget-based imagery, and "translational" research, has turned the best and brightest away from the lab bench and toward the spreadsheet.

It's occurred to me that the Folkman generation lived out in full the predictions made by Claude Bernard (1813–1878) for the future of medical science in his 1865 *Introduction à l'étude de la médecine expérimentale*.[4] At the time of its appearance, Bernard had become a Professor of Physiology at the prestigious Collège de France. In one brief decade (between 1848 and 1857), he had discovered the world of the internal milieu, had entered the heart of a dog by means of a catheter, and had established that blood glucose was derived from liver glycogen. He defined precisely the toxic actions of curare and carbon monoxide, discovered that pancreatic secretions broke down ingested fats, and produced experimental diabetes by puncture of the fourth ventricle in rabbits.[5] He was also the only European of his day to take note of the physiological work of William S. Beaumont, the forefather of experimental biologists in America.[6] Considering that he was born in Beaujolais, I'd call Claude Bernard all of biology in one full bottle.

As the star of Bernard's reputation rose, the level of his laboratory accomplishments declined. In response to fallow times, instead of falling on his sword, he fell on the pen, and his efforts were splendidly rewarded. Bernard's *Introduction,* written in clear, spare prose, became appreciated as a masterpiece of French literature taught in all the *lycées*. It was also a manifesto that transformed Western medicine from an observational into an experimental science.

> *Medicine necessarily begins with clinics, since they determine and define the object of medicine. But for a man of science there is no separate science of medicine or physiology, there is only a science of life. . . . In my opinion, medicine does not end in hospitals, as is often believed, but merely begins there.*[7]

AN ORGANIZING PRINCIPLE IN BIOLOGY

Judah Folkman's science began in hospitals, was hatched in the lab, and ended up improving human health the world over. Before he was forty, Folkman had made two major contributions to medical science; he lived to see each withstand the test of time.

His best-known contribution, the proposal of angiogenesis (new blood vessel formation) and anti-angiogenesis as an organizing principle, has held sway in experimental biology for decades. PubMed lists 34,109 citations since Folkman's first paper in 1971[8] and "angiogenesis" has ranked among the top five key words in the *FASEB Journal* since we've kept records.[9] In 1971, he isolated the first tumor-derived angiogenic factor and proposed that tumors require new blood vessels to grow and multiply.[10] Tumor angiogenesis, per se, was not an entirely novel proposal. Eli Moschkowitz, of Mt. Sinai, was the first to raise the notion of "angiogenesis" in pathology[11] and Philippe Shubik of Oxford then proceeded to discover that "tumor angiogenesis" was due to a substance or substances that could pass a membrane filter to induce new blood vessels in the hamster cheek pouch model.[12]

Folkman went these observations one better. Almost immediately after he had isolated a crude "tumor angiogenesis factor" in February 1971, he spelled out the import of his work in November.[13] He went on to suggest that tumors not only provided growth substances for angiogenesis, but that cancers were *absolutely dependent* on these factors in order to survive and kill their hosts:

> *It seems appropriate to speculate that the inhibition of angiogenesis, i.e. anti-angiogenesis, may provide a form of cancer therapy worthy of serious exploration.*[14]

Today, angiogenesis antagonists and stimulators are not only directed at cancer, e.g., Avastin® (bevacizumab), but also at macular degeneration, e.g., Lucentis® (ranibizumab). They are already in trials for conditions that range from coronary disease to rheumatoid arthritis to endometriosis. A generation after Folkman's paper in the *Journal of Experimental Medicine* (*Merci, maître Bernard*, for the name), more than ten new cancer drugs are on the market, and more than 1.2 million patients worldwide are receiving anti-angiogenic therapy.[15] The extension of Folkman's concept from the treatment of cancer to the preservation of sight comes from making sense of pathology in the light of biology:

> *Angiogenesis—the process of new blood-vessel growth—has an essential role in development, reproduction and repair. . . . However, pathological angiogenesis occurs not only in tumour formation, but also in a range of non-neoplastic diseases that could be classed together as "angiogenesis-dependent diseases." By viewing the process of angiogenesis as an "organizing principle" in biology, intriguing insights into the molecular mechanisms of seemingly unrelated phenomena might be gained.*[16]

Folkman's heuristic proposal, that angiogenesis is an organizing principle in biology, qualifies as a true discovery, the requirement for which was also defined by Claude Bernard.

We generally call a new fact a discovery; but I think that the idea which flows from that fact is the true discovery.[17]

FROM KITCHEN TO BANQUET HALL

Bernard held an exalted view of biology—the science of life—and reassured the young investigator that there was light at the end of the laboratory tunnel:

If a comparison were required to express my idea of the science of life, I should say that it is a superb and dazzlingly lighted hall which may be reached only by passing through a long and ghastly kitchen.[18]

Judah Folkman's path to the dazzling hall was shorter than most, and the kitchens in which he labored were by no means ghastly. And while the story of angiogenesis is well appreciated, Folkman made an earlier discovery in the course of medical training that put him on the road to angiogenesis; it was based on the observations that tumors leak hormones slowly into the bloodstream. Although millions of women the world over use contraceptives based on Folkman's discovery, the story of its application is largely unknown.

Folkman was born in Cleveland, attended Ohio State University, and as an undergraduate joined the experimental physiology laboratory of Dr. Robert Zollinger, a professor of surgery and no mean scientist himself. Like Bernard, he was a student of normal and abnormal pancreatic secretion: he discovered that non–beta-islet-cell tumors secreted gastrin, thereby inducing intractable peptic ulcers, a condition now known as the Zollinger/Ellison syndrome.[19] It did not escape Folkman's notice that tumors secrete bioactive materials.

From Ohio State, Folkman was admitted to the Harvard Medical School where, owing to skills developed in Zollinger's animal laboratory, he was taken under the wing of Robert E. Gross, Chief of Surgery at Children's Hospital. Gross's team, interested in repairing cardiac defects in infants, soon appreciated the importance of controlling the heartbeat during surgery, and Folkman was present as the first internal and external pacemakers entered the operating room.

After Harvard, he began internship and residency at the Massachusetts General Hospital, but after two years was drafted by the Navy. In keeping with the fast track open to Harvard surgeons, he was assigned to the National Naval Medical Center. His first assignment there was to develop a blood substitute —"When I was drafted, I was upset at the disruption to my surgical training," he told an interviewer.[20] Yet, his most widely applied discovery was made in Bethesda, "It turned out to be a terrific blessing in disguise."[21]

The "terrific blessing" turned into a matrix of scientific, social, and philanthropic interaction, featuring the National Naval Medical Center, the Dow Corning company, the Population Council, an independent research center working at Rockefeller University, and Wyeth Pharmaceuticals. The result: implantable contraceptives that have now been used by fifteen million women in sixty countries in every corner of the globe. The development was due, in part, to the "doctor draft" of the Korean war and its aftermath as some of the draftees found their way into labs at Walter Reed, the Bethesda Naval Hospitals, and the NIH. It was the dawn of the golden age at the Bethesda campus.

THE ENDOCRINE PACEMAKER

At the National Naval Medical Center, Folkman and Edmunds tried to form an "Endocrine Pacemaker for Complete Heart Block"[22] by implanting raw suspensions of cardiac stimulants such as triiodothyronine directly into dog hearts. They were out to relieve heart block biologically, by implanting tissue extracts or pure hormones instead of the electrical devices Folkman had studied at Harvard. Looking for ways to mimic the geometry of real endocrine glands, and to avoid grinding up thyroid tablets before they were injected into the dog's heart, Folkman and David Long sought to shield the active hormone with a number of synthetic and natural polymers, among them the biocompatible polymer Silastic® from Dow Corning. To label the injection site they enclosed a lipophilic dye, Trypan blue. To their surprise, when they opened the chest a few days later, they found that the Silastic® implants were absolutely pale: the dye had slowly leaked out.

It turned out that other drugs, dyes, and anesthetics could also be delivered in this fashion: Zollinger's student had finally found a way to leach bioactive substances into the circulation.[23] Years later, the concept was patented and assigned to Dow Corning (Midland, Michigan). Sheldon Segal, then responsible for research at the Population Council, described the next steps:

> *I learned about the diffusion of vital dye from Silastic® in 1966 while hosting a Dow-Corning representative (Silas Brady) in the Rockefeller University faculty dining room overlooking the East River. . . . He told me that Judah, also making use of the bio-compatibility feature, had coated experimental pacers, to prevent fibrosis when implanted in heart muscle . . . but, alas, discovered to his surprise that the dye had diffused out of the Silastic®. That was Judah's Eureka! moment. Mine was when I heard the story from Silas.*[24]

Segal's moment was also Folkman's: they had developed *the* endocrine pacemaker. Segal expanded on that moment by testing the release rate of steroid hormones from Silastic®. He found that synthetic progestins could be used in a system small enough to be a biocompatible contraceptive implant that could be inserted in the skin to have a long-acting life span.[25] After acquiring access to an appropriate contraceptive progestin (levonorgestrel) from Wyeth in 1968, Segal explained the council's work to Folkman, who immediately agreed to waive royalty rights for any product that might come out of the council's work and supported its cause with Dow Corning. Ultimately, the council received the waivers necessary to justify the major investment that would be required. Since its first launch in Europe in 1983, the popularity of Norplant®, or an authorized version has been steady. According to the UN Population Division's latest report, use of Norplant® has grown from five million worldwide users in 1990, to ten million in 2001, and fifteen million by the end of 2006. An additional three million women use a later version (Jadelle®).[26]

BACK TO THE LAB

These days, smart, well-motivated young docs in training are persuaded by their local sachems to join "Centers of Excellence in Health Care Delivery" or large multicentric "wellness" surveys. We could do worse than to ask the best and brightest of these to spend some time in labs like those of Bethesda Naval, Walter Reed, or the NIH—or simply to get back to the nearest bench in a lab. It would benefit not only experimental medicine and biology, but, as the career of Judah Folkman shows, humankind as well.

Claude Bernard would be on board:

> *I tell those whose path leads them toward theory or toward pure science, never to lose sight of the medical problem, which is to preserve health and cure disease. I tell those whose career, on the contrary, guides them toward practice, never to forget that if theory is meant to enlighten practice, practice in turn should be of use to science. . . . Experimental scientific medicine will thus become the achievement of us all; and every one of us will make his own useful contribution.*[27]

11. Apply Directly to the Forehead: Holmes, Zola, and Hennapecia

Dr. Oliver Wendell Holmes
(1809–1894)

Nana (Emile Zola).
Stamp issued by *La Poste* (2003)

There is nothing men will not do, there is nothing they have not done, to recover their health. . . . They have submitted to be half-drowned in water, and half-choked with gases, to be buried up to their chins in earth, to be seared with hot irons like galley-slaves, to be crimped with knives, like cod-fish, to have needles thrust into their flesh, and bonfires kindled on their skin, to swallow all sorts of abominations, and to pay for all this, as if to be singed and scalded were costly privilege, as if blisters were a blessing, and leeches were a luxury.

—Dr. OLIVER WENDELL HOLMES (1871)[1]

BRANDED BY HENNA

WERE DR. HOLMES TO OBSERVE BODILY MISCHIEF TODAY, HE'D STILL find needles thrust without cause into flesh and bonfires needlessly kindled on the skin. But, nowadays, the injuries are far less likely to be inflicted on the sick in search of health than on the vain in search of fashion. Botox bruises the foreheads of matrons, collagen scars the lips of barflies. Steel grommets hang from the navels of nymphets, bolts pierce the lips of perps. Perhaps the broadest practice, however, is the application of henna directly to hair and skin. This global assault has produced rock concerts that resemble the coming of age in Samoa and turned South Beach into the South Pacific. Warriors of the NFL sport body tattoos that put Papua to shame, while trendy folk in SoHo flaunt the umbilical Baroque. If the Belle Epoque was the Age of Gold, ours has become the Age of Tool and Dye.

Yet the medical literature documents that neither body piercing nor henna is all that safe.[2] Injuries provoked by cosmetic intrusion spare no age, no gender, no color, no class. Even the very young fall victim, as in a recent news item headlined SCARRED CHILDREN:

> *Michelle Lolk, of River Edge, took her 6-year-old daughter and 8-year-old son to a tattoo shop this past summer for their first-ever temporary tattoos. Young Ethan got a cross on his arm. His sister, Olivia, got a dolphin on her belly. A day later, Olivia complained of severe pain. "It looked like she was branded with a poker," Lolk said.[3]*

Skin branding of this sort (bonfires kindled on their skin, as Dr. Holmes might say) is due to acute contact dermatitis induced by henna's active agent, lawsone (2-hydroxy-1,4-naphthoquinone) and an added ingredient, PPD (para-phenylenediamine). Henna itself is a shrub (*Lawsonia inermis*, or Egyptian privet) cultivated in India, Sri Lanka, and much of North Africa. The dried leaves are mixed with various solvents and applied directly to the skin or hair. PPD is often added to red henna powder to produce the "black henna" preferred for tattoos.[4] But PPD also renders the mixture more allergenic and sometimes virulently toxic: à la the 1996 *Lancet* report "A Woman Who Collapsed after Painting Her Soles."[5] Temporary henna tattoos—of the sort applied at rock concerts and kiddie festivals—are intended to persist for only a few weeks but the incidence of acute inflammation, permanent scarring, and keloid formation has become epidemic in the last decade and a half.[6] PubMed lists only three reports of reactions to henna tattoo in the two decades between 1975 and 1995—but 259 papers since 1995! A number of these cases were caused by henna without PPD.[7] Hair dyes come in all sorts of proprietary formulations: a recent study from Korea reported that of 15 henna samples tested, PPD was present in three, nickel in eleven, and cobalt in four.[8]

Henna has been recognized as an occupational hazard in hair-dressing salons;[9] at various doses the dye induces hemolytic anemia in lab animals and humans alike;[10] oral intake of henna produces an acute inflammation of the colon.[11] In cell culture assays, lawsone causes cell death and cell cycle arrest in the S phase.[12] As might be expected for a redox-dependent naphthoquinine, individuals who lack oxidant defenses on a heritable basis, such as those with G-6 PD deficiency, are particularly at risk for Heinz-body hemolytic anemia.[13]

In response to injuries caused by "temporary tattoos," the FDA last year issued a warning against the import of henna preparations containing PPD, but explained that it was powerless to supervise ingredients in "cosmetic samples and products used exclusively by professionals—for example, for application at a salon, or a booth at a fair or boardwalk."[14]

So much for the kiddie trade!

HENNAPECIA OBSERVED

While the young are apt to apply henna directly to the skin, folks of a certain age mainly use henna to color their hair. The practice has been common for centuries in every corner of the world. Recently, however, France seems to have swept the honors for turning henna into art, as anyone can attest who has strolled through Paris. When the weather turns balmy, the streets are alive in a blaze of henna, offering coiffures in orange, auburn, red, and crimson.

But there may be a real downside to this display of vegetal finery. Over the years, on Parisian boulevards, in theaters, concert halls, cafés, and flea markets, I've observed a peculiar pattern of baldness in the French henna crowd. Women with hennaed hair, if of a certain age, seem almost uniformly to suffer from drastic, central alopecia (hair loss) quite evident at the back of their scalp, and quite noticeable in areas where their hair is parted. As a rheumatologist, I was struck by the difference between what one might call "hennapecia" and the commonly observed hair loss in patients with systemic lupus erythematosus. Nor is the pattern of hennapecia like that of ordinary female aging with its "increased thinning over the frontal/parietal scalp, greater density over the occipital scalp, retention of the frontal hairline, and the presence of miniaturized hairs."[15]

Alas, the phenomenon is not limited to France. Although hennaed hair is less common in the U.S., the same pattern seems to rear its ugly head, so to speak. Last fall, at the Pier Antiques Show in New York, where henna is also much in evidence, I observed consecutively forty-two women (approximate age over forty-five) with hair overtly dyed with henna. Twenty-nine had clear signs of hennapecia. As control, I observed thirty-six "blonde" women, presumably due to peroxide, of the same age: only five had similar areas of hair loss. In both groups, the incidence of exposed, undyed roots was pretty much the same.

Ever since Lewis Thomas and I consistently produced hair loss in rabbits given excess doses of vitamin A,[16] I have been intrigued by alopecia induced in lab animals and humans by agents of similar chemical structure. These have been associated with redox-induced changes in the hair growth cycle.[17] Lawsone and PPD are clearly involved in redox cycling and there are good reasons to believe that oxidative stress is involved both in graying and hair loss. This sequence was worked out by Arck and colleagues in the *FASEB Journal* in 2006. Entitled "Towards a 'Free Radical Theory of Graying,'" their paper concluded that "oxidative stress is high in hair follicle melanocytes and leads to their selective premature aging and apoptosis."[18]

Whatever the cause of hennapecia, it cannot be due to acute inflammation or contact dermatitis: the many scalps I've observed, albeit at a distance, seem to have been uninflamed. It is, of course, entirely possible that women prone to one or another type of genetic hair loss have a unique recourse to henna, but my guess would be that the striking correlation between henna and hair loss puts the onus of alopecia on the dye and/or its additives.

It's clear that to settle the point, we need experiment, not simple observation.

ZOLA'S EXPERIMENT: *NANA*

The distinction between experiment and observation was spelled out for the general reader by Emile Zola (1840–1902), and recently brought to mind by a lurid postage stamp issued by the French Postal Service in 2003 (illustration above). Few of the eager philatelists who snapped it up at first issue knew that the image was a poor caricature of a Manet portrait of Lucie Delabigne, the red-headed courtesan who became Zola's fictional Nana. Fewer still might have known that they were collecting a piece of scientific, as well as social, history. We can thank Claude Bernard, a founder of modern experimental medicine, for indirectly giving us Nana as well.

Emile Zola's 1880 novel *Nana* was a landmark of naturalist fiction, and overtly based on methods spelled out in Claude Bernard's *Introduction à la médicine expérimentale* (1865). In his *Le Roman expérimental* (also published in 1880), Zola declared that

> Claude Bernard . . . explains the differences which exist between the sciences of observation and the sciences of experiment. He concludes, finally, that experiment is but provoked observation. All experimental reasoning is based on doubt, for the experimentalist should have no preconceived idea, in the face of nature, and

should always retain his liberty of thought. . . . The essence of the higher organism is set in an internal and perfected environment [inherited characteristics] endowed with constant physico-chemical properties exactly like the external environment; hence there is an absolute determinism in the existing conditions of natural phenomena.[19]

Zola assigned Nana a constant internal milieu, the "inherited characteristics" of a feral, manipulative courtesan. He then exposed his heroine to a variety of external stimuli (journalists, bankers, actors, gentry, tycoons), varied the strength of the buffer (theaters, garrets, hotels, mansions), and changed the ambient oxygen tension (age, war, disease, death). He then recorded the results of these true-life interactions in the laboratory notebook of his naturalist novel.

Zola introduces Nana at time zero of his experiment. In her first appearance on stage, she is clad only in a diaphanous, see-through gown:

Applause burst forth on all sides. In the twinkling of an eye she had turned on her heel and was going up the stage, presenting the nape of her neck to the spectators' gaze, a neck where the golden red hair showed like some animal's fell. Then the plaudits became frantic.[20]

The experiment proceeds, her nature (feral, manipulative) remains constant; only the stimuli and buffers change; the lovers and venues vary. At the book's end, after she has been exposed to a repertoire of environmental stress (age, disease, death), Nana dies of syphilis. A bellicose crowd marches under her windows at the Grand Hotel screaming "On to Berlin" as the Franco-Prussian war begins. Nana fades away, along with the Second Empire of Napoleon III: "On the bed lay stretched a gray mass, but only the ruddy chignon was distinguishable and a pale blotch which might be the face."[21]

Zola described the faith of a nineteenth-century realist, in a passage that remains as pertinent to experimental biology as to the art of the novelist:

The novelist is equally an observer and an experimentalist. The observer in him gives the facts as he has observed them . . . then the experimentalist appears and introduces an experiment, that is to say, sets his characters going in a certain story so as to show . . . the machinery of his intellectual and sensory manifestations, under the influences of heredity and environment, such as physiology shall give them to us.[22]

Perhaps that's the machinery novels would still be exposing had novelists remained in touch with experimental science. And perhaps the FDA might send people to watch the influence of heredity and the environment on henna applied directly to the forehead.

12. Elizabeth Blackwell Breaks the Bonds

Albert Lewis Sayre at Bellevue Hospital
Medical College. "Spinal Traction" (1876)

Women's Medical College of New York Infirmary
Leslie's Illustrated Newspaper, April 16, 1876

You ask me what I did, and what can be done as a lady. I entered the Maternité, dissected at l'Ecole des Beaux-Arts alone, employed a répétiteur who drilled me in anatomy and smuggled me into the dead-house of La Charité at great risk of detection, where I operated on the cadaver. I once made the rounds of his wards in the Hôtel-Dieu with Roux, heard his lectures, and saw his operations. I attended lectures at the Collège of France and Jardin des Plantes.

It is my impression, for I ought only to put it in that modest form, that the ruling class in America is less humane, more addicted to money-getting and party spirit [than Europe]; and that reform ideas in America are much more talked of, but less acted on.

—ELIZABETH BLACKWELL, letters to sister Emily (1850–54)[1]

PREJUDICE IS MORE VIOLENT
THE BLINDER IT IS

THE IMAGES ABOVE DOCUMENT A SEA CHANGE IN AMERICAN MEDICINE. After Elizabeth Blackwell broke the bonds, women were no longer there simply to be acted upon (left), but could themselves take action (right). The *New York Times* noted the critical event in adjacent news reports on March 1,

1867. The evening before, at Steinway Hall on 14th Street, Dr. Albert Lewis Sayre had presided over the graduation ceremonies of Bellevue Hospital Medical College at which 140 gowned men marched down the aisle to the music of Wagner.[2] A founder of the school, and the first Professor of Orthopedic Surgery in the U.S., Dr. Sayre was famous not only for his "suspension" treatment of spinal deformities but also for a textbook illustrated by provocative dorsal views of female patients. On the same evening, and only three blocks away at the New-York Historical Society on 11th Street, Dr. Elizabeth Blackwell addressed an audience of men and women on the subject of "The Medical Education of Women." The Women's Medical School that she founded with her sister had just been chartered, and she appealed for funds to permit women to gain "a thorough knowledge of the science of medicine."[3] Scientific knowledge of the body, she pleaded, could only be obtained by practical, hands-on work, as in the revolutionary practice of learning anatomy by dissection. *Leslie's Weekly* the shocking news that a woman could become a doctor over the dead body of a man (see illustration, above).[4]

The founding of a medical school for women had its origins in 1854 when Elizabeth Blackwell, described by *Lancet* as "the first woman medical graduate in the modern meaning of the phrase,"[5] arrived in New York after clinical work in Paris and London. In Paris she had been exposed to the bracing notion of Lamarckian evolution at the Jardin des Plantes and to the experimental medicine of Magendie and Claude Bernard at the Collège de France. In London she acquired the fervor of sanitary and social reform, and became a lifelong devotee of Florence Nightingale.

America in 1854 was not ready for Elizabeth Blackwell, nor for other women of her stripe and, alas, our country remains unready for them today. Although many women have found their place in the public sphere, some near, if not at, the very top, our country remains addicted to customs Blackwell called "money-getting and party spirit" (read zealotry). Two current items remind us of the prebellum country that Blackwell encountered on her return, an America in which money-getting trumps the facts of science and zealots wave the flag of bigotry.

ITEM: THE REPUBLICAN CANDIDATES DEBATE

MR. VANDEHEI: I'm curious, is there anybody on the stage that does not agree— believe in evolution? (Senator Brownback, Mr. Huckabee, Representative Tancredo raise their hands.)   —*New York Times*, May 3, 2007[6]

ITEM: RELIGIOUS LEADERS RIP HATE CRIME BILL

A hate-crimes bill passed Thursday by the House, extending coverage to people victimized because of sexual orientation, gender identity or disability, is attracting opposition from an unusual . . . coalition of evangelical, fundamentalist and black religious leaders that is mounting a furious assault on the bill, airing television ads and mobilizing members to stop its progress. And President Bush has said he may veto the measure.

—*Chicago Tribune,* May 5, 2007[7]

A HUNDRED YEARS HENCE

Born in England to Samuel Blackwell, a well-off sugar refiner and dissident lay preacher, Elizabeth Blackwell was brought to Cincinnati in 1832. Her education was peripatetic, in circles where abolitionist politics and Transcendental values held sway. She received medical tutorials in the private practices of Philadelphia doctors, but despite her thorough preparation in textbook anatomy and a solid educational record, she was refused admission by seventeen medical faculties in the United States. However, there was a small college in Geneva, New York, that granted the degree of Doctor of Medicine, provided lectures were attended for two years and a thesis was written. In those pre-Flexnerian days, Geneva's requirements for the M.D. degree were par for the course, as it were.[8]

Blackwell matriculated in November 1847, was more or less well received by town and gown, and performed splendidly in all classwork, especially therapeutics. In the summer of 1848, she undertook clinical instruction at the Philadelphia Hospital, where "ship fever" (epidemic typhus) had broken out among Irish immigrants. Blackwell carefully recorded its spread from case to case and recommended prevention by light, air, and the washing of hands with soap and water. This exercise in clinical epidemiology became her doctoral thesis, a work almost as persuasive as that of Oliver Wendell Holmes on puerperal fever (1843). Indeed, Blackwell's thesis relied heavily on the work of Holmes's teacher, Professor P.C.A. Louis of Paris, who first distinguished typhoid from typhus. By February 1849, Blackwell's thesis had been published in the *Buffalo Medical Journal and Monthly Review,*[9] and all that remained for her doctorate was to graduate with distinction.

CLAUDE BERNARD,
A DISTINGUISHED YOUNG INQUIRER

The degree won, Blackwell determined to obtain the best clinical training possible. Since in those days young American doctors properly regarded Paris as the center of clinical science, Blackwell sailed off to the City of Light.

She arrived with an introduction from her Philadelphia preceptors to none other than Professor P.C.A. Louis himself, "then at the height of his reputation." She felt instinctively that his visit was one of inspection and passed with flying colors. Thanks to Louis's intervention, she was admitted in autumn of 1849 for a six-month course at the Maternité lying-in hospital. In that conventlike atmosphere her most intellectually stimulating companion was the intern, M. Hippolyte Blot. (In later years Blot went on to a professorship at the Maternité, having discovered the relationship between eclampsia and kidney disease.[10])

Intern Blot and pupil Blackwell exchanged lessons in English and histology, spending hours over the young man's microscope. Blot taught Blackwell that the human red blood cell was biconcave, streaking a drop of blood on a slide to show "that what appeared to be a central spot in each globule was owing to the convexity not being in focus, and it disappeared when the focus was a little lengthened." In the Paris of 1849 experimental medicine was everywhere in the air. Hippolyte Blot told Blackwell of his friend,

Claude Bernard, a distinguished young inquirer, who is now, he thinks on the eve of a discovery that will immortalize him . . . of the power which the liver has of secreting sugar in a normal state when animals are fed on certain substances which can be so converted; also of the curious experiment by which a dog was made, in his presence, to secrete albuminous or diabetic urine.[11]

BLACKWELL AND THE GONOCOCCUS

Attending to her patients at the Maternité, a grave accident befell Elizabeth: "in the dark early morning, whilst syringing the eye of one of my tiny patients for purulent ophthalmia," some of the water spurted into her left eye. By nightfall of November 4, the eye had become swollen, and by the next morning, the lids were "closely adherent from suppuration." The diagnosis of purulent ophthalmia, the dreaded venereal disease of newborns and those who attended them, was made and the twenty-eight-year-old Blackwell was placed in the student infirmary. The disease is caused by the gonococcus, is due to chronic gonorrheal infection of the female reproductive tract, and was part of the load borne by the prostitutes and working women who gave birth in the public hospitals of Paris. The bacteriologic revolution has all but eliminated it, but Albert Neisser did not discover the microbe until 1879, and it was not until 1884 that Carl Credé made it clear that eyedrops of 1% silver nitrate on the lids of newborns were an effective prophylactic. Thanks to rigorous maternal health laws, by the turn of the century prophylaxis had pretty much eliminated neonatal ophthalmia from advanced countries; but today resistant strains are making an unfortunate comeback in many parts of the globe.[12]

Elizabeth Blackwell was treated by the accepted methods of the day: cauterization of the lids, leeches to the temple, cold compresses, ointment of belladonna, opium to the forehead, purgatives, and footbaths. Dr. Blot came in every two hours, day and night, to tend the eye. But despite his efforts, after three days it became obvious to her doctors that the eye was hopelessly infected.

> *Ah! how dreadful it was to find the daylight gradually fading as my kind doctor bent over me and removed with an exquisite delicacy of touch the films that had formed over the pupil! I could see him for a moment clearly, but the sight soon vanished, and the eye was left in darkness.*[13]

She lay in bed with both eyes closed for three weeks, but then the right eye gradually began to open. Soon she could begin to perform little tasks for herself: she assured an uncle in England that she could write without difficulty, read a little, and hoped to return to her studies. She remained permanently blind in one eye and because of this handicap disqualified herself from surgery or obstetrics as a career.

Blackwell next applied to St. Bartholomew's Hospital in London, then as now perhaps the strongest teaching hospital of the city. The illustrious Sir James Paget endorsed her admission as a student "in the wards and other departments of the hospital" and on May 14, 1850, she was accepted at Bart's. Once on the wards, she soon spotted the difference between the medicine of Paris and London at mid-century:

> *I do not find so active a spirit of investigation in the English professors as in the French. In Paris this spirit pervaded young and old, and gave a wonderful fascination to the study of medicine, which even I, standing on the threshold, strongly felt.*

But overall, her London hours were instructive, she made many new female friends, some socially prominent, among them Miss Nightingale. She also encountered more than the expected rebuffs. Her mentor gave her sound advice, which elicited a passionate response:

> *Mr. Paget who is very cordial, tells me that I shall have to encounter much more prejudice from ladies than from gentlemen in my course. I am prepared for this. Prejudice is more violent the blinder it is . . . but a work of the ages cannot be hindered by individual feeling. A hundred years hence women will not be what they are now.*[14]

WOMEN WILL NOT BE
WHAT THEY ARE NOW

Her experiences in Paris and London made her anxious to begin on her own in America. In November 1850 she wrote of her future plans to her sister Emily, who had decided to follow in her sister's footsteps: "I shall commence as soon as possible building a hospital in which I can experiment." On her return to New York she was too poor to realize the dream of building an experimental hospital. "If I were rich," she had written her sister, "I would not begin private practice, but would only experiment. As however I am poor, I have no choice." She set up a general practice and spent cold winters and steaming summers in the city trudging the pavements with her black bag. Her early years as this country's first woman doctor of medicine were not encouraging. She confessed deep unhappiness: "I had no medical companionship, the profession stood aloof, and society was distrustful of the innovation. Insolent letters occasionally came by post, and my pecuniary position was a source of constant anxiety."[15]

It was impossible to rent an office, the term "female physician" having been preempted by ill-trained abortionists, and she went into debt by buying a house on East 15th Street. She worked in the attic and basement, renting out the remainder of the house. Her isolation prompted her to adopt a seven-year-old orphan, Katharine Barry, and this young child grew into a lifelong companion, friend, and housekeeper.

Slowly, Elizabeth Blackwell began to attract support from the New York Quaker community, and by 1854 she had opened a one-room dispensary on the Lower East Side. The dispensary treated over two hundred women in its first year. By 1856, Elizabeth was reunited with her sister who had received medical training in Europe after an M.D. from Western Reserve. With the help of progressive philanthropists like their good friend Horace Greely, the Blackwells established the New York Infirmary for Women and Children in 1857 on 64 Bleecker Street. They successfully overcame each of the social objections of the time: that female doctors would require police protection on their rounds; that only male resident physicians could control the patients; that "classes and persons" might be admitted whom "it would be an insult to treat" (i.e., beggars and prostitutes); that signatures on death certificates might be invalid (the legal rights of women in the presuffrage era were fragile); that the male trustees might be held responsible for any "accidents"—and that in any case no one would supply women with enough money to support such an unpopular effort.

With Emily now in charge of a going concern, Elizabeth traveled back to England and became the first woman to be registered as a physician in the UK. She studied programs of maternal hygiene, looked over public-health programs for women and children, and toyed with the notion of spending the rest of her life working in a country hospital together with Florence Nightingale, with whom she had formed an intense personal relationship.[16]

KNOWLEDGE NOT SYMPATHY

When Elizabeth returned once more to New York in 1860, the sisters enlarged the infirmary, added new staff, and put in place the preventive measures of the Sanitarian revolution. The Civil War fully engaged their abolitionist spirit. On the day after Fort Sumter was fired on, the Blackwells helped to found the National Sanitary Aid Society (in turn, the Sanitary Commission), a major service to public health in the Union cause. With war over, Elizabeth's dream was realized: a hospital in which to experiment. In 1867–68 the sisters founded the Women's Medical College of New York Infirmary, which by 1899 had graduated 394 women doctors! The laboratories for instruction in both basic and applied sciences were among the most up-to-date in the country, and the three-year curriculum exceeded in rigor much of what passed for medical education in this country. Elizabeth Blackwell became the first Professor of Hygiene and it was due to her efforts that hands-on science—anatomy, histology, physiology—came first:

> *It is observation and comprehension, not sympathy, which will discover the kind of disease. It is knowledge, not sympathy, which can administer the right medicine; and though warm sympathetic natures, with knowledge, would make the best of all physicians, without sound scientific knowledge, they would be most unreliable and dangerous guides.*[17]

On the one hand, the social causes to which the Blackwells devoted themselves have by and large prevailed: the medical education of women advanced, the Union preserved, sanitation promoted, infections curbed, child and maternal health protected by the state, and so forth. On the other hand, we have a long way to go. The personal lives of Elizabeth and Emily Blackwell remained private and monogamous. Aside from that passionate episode with Florence Nightingale, Elizabeth spent all of her life—the last thirty years in seaside retirement—with her adopted daughter/friend Kitty Barry. Emily and her lifelong companion, Dr. Elizabeth Cushier, spent twenty-eight happy years together in a Gramercy Park brownstone and on the coast of Maine.

Sadly, these days, perhaps the only place Elizabeth Blackwell and Florence Nightingale could live together in legal peace would be in Massachusetts. We are still some distance from realizing her fondest hope, written in reply to an invitation from the Convention for Women's Rights in Worcester, Massachusetts (1850):

> *The great object of education has nothing to do with woman's rights or man's rights, but with the development of the human soul and body. My great dream is of a grand moral reform society, a wide movement . . . combined that it could be brought to bear on any outrage or prominent evil.*

13. Chronic Lyme Disease and Medically Unexplained Syndromes

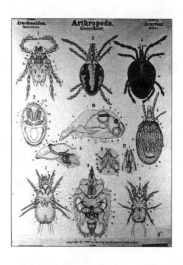

Rudolph Leuckart, *Arthropoda* wallchart.
From MBLWHOI Library, used with permission[1]

DIAGNOSIS, n. A physician's forecast of the disease by the patient's pulse and purse.
—AMBROSE BIERCE, *The Devil's Dictionary,* 1911[2]

ANTI-TRUST AND THE TICK

LYME DISEASE, A TICK-BORNE BORRELIOSIS, MADE HEADLINES IN 2006 IN the business world: LYME DISEASE GUIDELINES FOCUS OF ANTITRUST PROBE.[3] The attorney general of Connecticut threatened to invoke antitrust laws against the Infectious Diseases Society of America, a scholarly society of physicians who had issued a set of guidelines for the treatment of Lyme disease. The guidelines discouraged use of intravenous antibiotic therapy for the late, non-specific, neurological symptoms of "chronic Lyme disease." The infectious disease panel of experts noted that

> *In many patients, post-treatment symptoms appear to be more related to the aches and pains of daily living rather than to either Lyme disease or a tickborne coinfection. Put simply, there is a relatively high frequency of the same kinds of symptoms in "healthy" people.*[4]

But a few self-proclaimed experts and a vociferous group of Lyme disease advocacy groups argued that medical science has it wrong and that the establishment is denying treatment to desperate patients. The advocates insisted that only vigorous, intravenous antibiotic therapy can relieve "Lyme victims" of their chronic pain, fatigue, and neurologic complaints. Diane Blanchard, of the Connecticut-based organization Time for Lyme, complained, "These guidelines are becoming the de facto standard of care and that is not OK. We are all guinea pigs at this point. Why would anyone think they have all the answers? It's not right."[5] There's money at stake: since the guidelines have been endorsed by the Centers for Disease Control and Prevention, insurers will be unlikely to pay for antibiotics given by vein.

The attorney general responded to the Lyme advocacy position in businesslike fashion, holding, "These guidelines were set by a panel that essentially locked out competing points of view. Presumably, the IDSA is a nonprofit making organization, but such organizations can still be used for anticompetitive purposes."[6]

He invoked the antitrust laws, and the case is still open. In fact, the panel of medical scientists had already looked into "competing points of view." Citing properly controlled studies, the panel found no support for the use of long-term, potentially dangerous antibiotic therapy in the absence of objective physical signs or lab abnormalities. They noted that population-based surveillance in the United States indicated a mean of 6.1 self-reported unhealthy days during the preceding month.[7] They concluded,

> *Thus, the presence of arthralgia, myalgia, fatigue, and other subjective symptoms after treatment for Lyme disease must be evaluated in the context of "background" complaints in a significant proportion of individuals.*[8]

This sort of "background" is illustrated by a case report written by the complainant herself in a letter to the editor of a Canadian newspaper:

> *I had a flu-like illness with severe fatigue, muscle pain, fever and chills. . . . It lingered on and on for weeks, then months, then years. There was always chronic pain and debilitating fatigue that made it impossible to keep working. The diagnosis was fibromyalgia and later chronic fatigue syndrome. . . . During those years, I suffered from severe joint and muscle pain and very often extreme fatigue. Later, after having a very strained immune system, I developed allergies and multiple chemical sensitivities. Today, 22 years later, after having a blood test sent to a California lab along with a history of my symptoms, Lyme disease has been confirmed.*[9]

There is, of course, no way for a California lab to have made the diagnosis of Lyme disease on the basis of a blood test twenty-two years after a flu-like syndrome, even when provided with a "history of my symptoms." "No tickee, no washee," as they used to say before *Ixodes dammini* (the tick) or *borellia burgdorferi* (the microbe) were on the map.

FROM PARALYSIS TO FATIGUE

Alas, diagnoses such as chronic Lyme disease, based on the pulse of the time and the purse of the patient, are far too common these days. Groups of patients and advocates march against medical science under the banners of "chronic fatigue syndrome," "myalgic encephalitis," "irritable bowel syndrome," "total chemical allergy," etc. There is no question that patients suffer —and often terribly—from conditions to which these labels have been given. There is also no question that their disability is real. Skeptics worry, however, that the hallmark of these "diseases" is that diagnosis requires the complete absence of objective physical or biochemical derangement. They wonder whether such patients are not really victims of a complex set of socially and medically constructed diseases—much as the "railway spine," "chronic appendicitis," or "female hysteria" favored by nineteenth-century clinicians. These doubts are summarized in Edward Shorter's *From Paralysis to Fatigue: A History of Psychosomatic Illness in the Modern Era.*

> *Although the amplification of normal bodily symptoms and phobias about disease have existed in all times and places, it is this delusional clinging to the belief in a given illness, that marks the last decades of the twentieth century.*[10]

Shorter is persuaded that social templates shape medical fashion and that medical fashion shapes the symptoms that patients select. Those symptoms— such as fatigue, weakness, tinglings, insomnia, etc.—could, of course, be produced by organic disease; that's exactly why they tug so hard at our diagnostic sleeve. The victim of chronic Lyme disease is in very real pain—but of the mind—and the mind chooses symptoms that will be taken as evidence of physical disease and that will win the patient an appropriate response.

> *Thus most of the symptoms . . . have always been known to Western society, although they have occurred at different times with different frequencies: Society does not invent symptoms; it retrieves them from the symptom pool.*[11]

MEDICALLY UNEXPLAINED SYNDROMES

The British psychiatrist Simon Wessely of King's College, London, explains that most medical specialties define unexplained syndromes in the technical terms of their own specialty. Presented with the same cluster of symptoms by a patient, a rheumatologist will call "fibromyalgia" what a gastroenterologist would diagnose as "irritable bowel syndrome," while a neurologist might come up with "chronic fatigue syndrome" and a dabbler in infectious disease would label the syndrome "chronic Lyme disease." Wessely provides convincing evidence that none of these monickers describes a unique clinical entity. Indeed, each syndrome shares much with all the others: muscle weakness, arthralgias, and overall fatigue, the repertoire of symptoms in "chronic Lyme disease." Skip the label, Wessely advises, it's better to describe these conditions, honestly, as "medically unexplained syndromes."[12]

Others have debated whether it helps patients to have labels such as "fibromyalgia," "chronic Lyme disease," "total chemical allergy," "chronic fatigue syndrome (CFS)" pinned on their ills. Henninsen, et al. have suggested that

> *The answer to the question of "to label or not to label" may turn out to depend not on the label, but on what that label implies. It is acceptable and often beneficial to make diagnoses such as CFS, provided that this is the beginning, and not the end, of the therapeutic encounter.*[13]

That encounter does not include intravenous antibiotic therapy for medically unexplained syndromes.

ZEITGEIST IS AS ZEITGEIST DOES

Chronic fatigue syndromes are also found in children; the condition might be called "Münchhausen's fatigue by proxy." A British study found that children are most likely to develop CFS in the autumn term when they start secondary school: seventy-six percent of children in one study developed CFS between September and December. On average, the children in the study, who were otherwise healthy, were eleven years old when the illness began, coinciding with their move to secondary school.[14] Since infective diseases in childhood are far more common in kids from poor families it was noteworthy that most of the sufferers of childhood CFS were from "higher socio-economic" (i.e., rich) families. That seems to be true for most other medically unexplained physical syndromes, such as chronic Lyme disease. Indeed, the direct correlation between income and fatigue syndromes is an argument for the social-construction hypothesis versus the usual "infective" or "somatic" etiology of these troubling conditions.[15] Score one for the infectious disease experts: social construction is not amenable to intravenous antibiotic therapy. In vein, in vain for the unexplained.

Shorter explained the historical patterns into which those unexplained syndromes fall; patients tend to introject the bad dreams of their *zeitgeist*, paralysis in the old days of syphilis, fatigue in the era of AIDS.[16] He warns doctors not to regard "patients with 'somatoform' symptoms as bizarre objects but as individuals who enjoy the dignity that all disease confers." On the other hand, doctors tend to be impatient with those who come to them with inexplicable symptoms. No tick, no fever, no rash, no changes in the spinal fluid equals no disease, they feel. No wonder doctors are often at odds with their patients, who, in the words of Sarah Nettleton of York University, "just want permission to be ill":

> *Indeed, society does not readily grant permission to be ill in the absence of disease . . . an appreciation of the experience of such embodied doubt articulated by people who live with medically unexplained diseases may have a more general applicability to the analysis of social life under conditions of late modernity.*[17]

LIKE A HOLE IN THE HEAD

But patients with medically unexplained disease continue to suffer, each in the fashion of the day, each in search of the most radical remedy, be it cauteries or antibiotics. Perhaps the saddest response to pain *sans* disease was reported a while ago by Reuters:

> BRITON CURES FATIGUE BY DRILLING HOLE IN OWN HEAD
> February 22, 2000 LONDON (Reuters)—A British woman says she has cured her chronic fatigue by resorting to do-it-yourself brain surgery and drilling a hole in her own head. Heather Perry, 29, performed the ancient technique of trepanning, cutting away a section of the scalp and drilling into the skull.[18]

Perry tried to rid herself of her chronic fatigue syndrome by drilling a two-centimeter hole in order "to permit blood to flow more easily around the brain." But the operation went wrong when she drilled too far and penetrated the dura mata. British doctors had refused to help Perry with the ancient procedure, so she flew to an unnamed location in the United States, where she was given medical advice and then did it herself. She said the twenty-minute operation had improved her quality of life.

> *"I have no regrets. I was prone to occasional bouts of depression and felt something radical needed to be done," said Perry, who performed the operation under local anaesthetic in front of a mirror and a camera crew.* [What a segment for T.V.!

On to Oprah or Larry King.] *"I felt the effects immediately, I can't say they have been particularly dramatic but they are there. I generally feel better and there's definitely more mental clarity. I feel wonderful.*[19]

Trepanning for fatigue has as little support in science as the intravenous antibiotic therapy urged on the attorney general of Connecticut, but it does have the advantage of not adding to the expenses of Britain's National Health Service or American insurers. *The Devil's Dictionary* requires revision: nowadays diagnosis falls under the antitrust laws, while the pulse and the purse are those of the public.

14. Eugenics and the Immigrant: Rosalyn Yalow and Rita Levi-Montalcini

Rosalyn Yalow (Prix Nobel, 1977)   Rita Levi-Montalcini (Prix Nobel, 1986)

My father, Simon Sussman, was born on the Lower East Side of New York, the Melting Pot for Eastern European immigrants. . . . Since I could type, [I] obtained a part time position as a secretary to Dr. Rudolf Schoenheimer, a leading biochemist at Columbia University's College of Physicians and Surgeons (P&S). This position was supposed to provide an entrée for me into graduate courses, via the backdoor, but I had to agree to take stenography. —ROSALYN YALOW[1]

On 19 December 1946, Renato Dulbecco and I sailed from Genoa on board the Polish ship, the Sobieski, I headed for St. Louis and he for Bloomington. When the Statue of Liberty became visible against the sky of the port of New York . . . I felt as hundreds of thousands of refugees have felt, in the flight from recent as well as earlier persecutions upon arrival in New York Harbor. . . . My stay lasted thirty years. —RITA LEVI-MONTALCINI[2]

RECENT AND EARLIER PERSECUTIONS

THE BIOGRAPHIES OF ROSALYN YALOW AND RITA LEVI-MONTALCINI yield *prima facie* arguments for liberal immigration and visa policies. Yalow's story particularly illustrates how bigotry and eugenic notions led to the discovery of radioimmunoassay in the United States. Arrivals like those of Yalow's grandfather from Czarist Russia or Levi-Montalcini from Fascist Italy are perhaps only small episodes in the story of America's rise to pre-eminence in science. Indeed, other factors surely played larger roles: the GI Bill of Rights, Shannon's NIH, public access to higher education (Hunter College for Yalow), private philanthropy (Rockefeller, Hughes), etc. But we'd surely be many notches down in science had our borders been closed to arrivals from abroad.

Twenty years ago, its repute in apogee, the United States accounted for about 40 percent of the total number of reputable scientific papers published in the world, the European Union for 33 percent, and the Asia-Pacific region for 14 percent.[3]

Those days are over: the seats of American power have been usurped by fans of unreason, bible-thumpers who feel free to preach "creation science," "alternative medicine," "faith-based" social service, and blatant homophobia. In consequence, the standing of American science has been eroded. By 2004, the EU had moved into the lead with 38 percent of total scientific papers published worldwide, the United States had slipped to 33 percent, while the Asia-Pacific region moved up rapidly to become the source of 25 percent of all papers. It's hard to see how federal action to prevent flag-burning or gay marriage can address this issue.

Nor is it helpful to disguise bans on scientific exchange under the scoundrel's cloak of national security. "Scientists Denied U.S. Visa," the headlines scream,[4] while the president of Intel complains to the *Financial Times* that

> *America is experiencing a profound immigration crisis but it is not about the 11 million illegal immigrants currently exciting the press and politicians in Washington. The real crisis is that the U.S .is closing its doors to immigrants with degrees in science, maths and engineering.*[5]

Data on the U.S. work force in science can be used to make another argument for liberal immigration and visa policies. The 2000 census documents that, whereas Asians comprise only 4.1 percent of the total work force in the U.S., 14.7 percent of all U.S. life scientists are Asians![6] We can be grateful that, in deference to our wartime alliance with China, the Roosevelt administration in 1943 repealed the racist Chinese Exclusion Act, which had essentially excluded all Asians from the continental United States since 1881.

We have further reason to thank FDR when we examine publications in the life sciences today. In the March Issue of the *FASEB Journal,* one counts 41 articles (Research Communications and FJ Express) with 312 authors listed, an average of 7.6 per article. Of those authors, 98 (31 percent) had overtly

Asian surnames (Indian, Chinese, Japanese, etc.), split evenly between scientists working in U.S. labs and abroad. That squares with the 15 percent of Asians in the work force of life sciences overall in the U.S. (see above).

In the March 24, 2006, issue of the *Journal of Biological Chemistry* one counts 65 articles with 399 authors listed, an average of 6.1 authors per article. Of those authors, 188 (47 percent) had Asian surnames (Indian, Chinese, Japanese), with 83 (21 percent) working in U.S. labs and 105 in labs abroad (26 percent). Now let's compare those data with the years of the American apogee.

The March 25, 1986, issue of the *Journal of Biological Chemistry* contained 76 articles with 260 authors listed, a more modest average of 3.4 per article. Of those authors, 44 (16 percent) had Asian surnames, again evenly split between Asians working in U.S. labs and abroad. That's less than half the number of 2006!

That doubling of Asian contributions to American science between 1986 and 2006 is directly due to liberal visa and immigration policies. A generation before, these had brought Renato Dulbecco to Bloomington to work with Salvador Luria and led Rita Levi-Montalcini to discover nerve growth factor with Stanley Cohen in St. Louis rather than in Turin.

EUGENICS AND IMMUNOASSAY

Karl Pearson, FRS (*left*) (1857–1936) and Francis Galton, FRS (*right*) at age 81 (1822–1911)

Rosalyn Yalow's evocation of the Lower East Side of New York as "the Melting Pot for Eastern European immigrants" reminds one that entry of Eastern European Jews into Anglo-Saxon lands was as much of a political issue at the dawn of the twentieth century as Mexican immigration at the dawn of the twenty-first. In Britain, their exclusion was championed by two leaders of the eugenic movement, Francis Galton and his student Karl Pearson.

In 1925, Karl Pearson, together with Margaret Moul, published an extensive two-part analysis of "The Problem of Alien Immigration into Great Britain, Illustrated by an Examination of Russian and Polish Jewish Children."[7] The paper was the lead article in the *Annals of Eugenics* published by the Galton Institute (K. Pearson, ed.). By means of a detailed study, carried out before World War I, of over a thousand Jewish schoolchildren recently arrived in England from Eastern Europe, the authors attacked the problem of whether the intelligence of these immigrants differed from that of the native stock.

The study was a model of biometric detail: Pearson remains a scientist of repute, whose contributions to biostatistics have remained practically untarnished. Not only were the children he studied given the most modern tests of intelligence, but school records were examined, home visits made, and physical examinations performed. Control groups were found: English-born Jews and native Gentiles. Elaborate scoring systems were employed to evaluate such variables as size and income of family, rent paid, foci of infection, crowding, ventilation, mouth-breathing versus nose-breathing, "cleanliness," etc.

The authors directed their inquiry to an applied end, in keeping with the overall aims of the eugenics movement:

> We hold therefore that the problem of the admission of an alien Jewish population into Great Britain turns essentially on the answer that may be given to the question: Is their average intelligence so markedly superior to that of the native Gentile, that it compensates for their physique and habits certainly not being above (probably a good deal below) the average of those characters here?

Pearson and Moul found that, for all groups examined, there was no correlation between intelligence and any other variable such as cleanliness, mode of breathing, family size or income, foci of infection, and height-for-age. Consequently, they were led to this rather somber conclusion:

> . . . the argument of the present paper is that into a crowded country only the superior stocks should be allowed entrance, not the inferior stocks, in the hope—unjustified by any statistical inquiry—that they will rise to the average native level by living in a new atmosphere. The native level is not a product of the atmosphere, but of centuries of racial history, selection, hybridisation and extermination.

Extermination? As in roaches, one is tempted to ask? Be that as it may, the authors failed to note a curious anomaly among their data. All variables considered, there was a striking difference in "intelligence" between Jewish girls and boys, the latter being statistically more intelligent.

> *Namely, that with the Gentile children we have found only a slight difference between the boys and girls. Hence the intelligence of the Jewish girls being much below that of the Jewish boys, even if the latter equaled that of the Gentile boys, the Jewish girls would fall very seriously behind the Gentile girls.*

One must point out the genetic fallacy here. If conclusions from such data were possible, we could with some degree of confidence say that in Eastern Jews, by some unusual genetic aberration, intelligence was sex-linked, whereas in Gentiles this higher faculty was not.

In the event, arguments such as these directed the bulk of immigrants from Eastern Europe to the Lower East Side of New York rather than to Cheapside in London. We therefore have Karl Pearson and his fellow eugenicists to thank for their indirect gift to American science: permitting Rosalyn Yalow with Solomon Berson to develop radioimmunoassay at a Veterans Administration Hospital in the Bronx.

15. Science in the Middle East: Robert Koch and the Cholera War

Robert Koch (1843-1910) Memorial Stamp

ROCKETS AND MCATS

IN THE LAST DAYS OF JULY 2006, AS HEZBOLLAH ROCKETS LANDED OUTSIDE an operating theater at the Rambam Medical Center in Haifa, a surgeon continued with his operation and the website of Haifa University announced that "This week no exams will be held, neither will lectures be given."[1] On the same day, while a lighthouse near the American University of Beirut was hit by Israeli bombs, its provost left the country and the university posted the news that "We regret, for many reasons, being unable to administer the August 2006 MCAT examination in Beirut."[2]

That was the outlook for our young colleagues in the Middle East at the time: canceled biochemistry exams in Haifa, pre-meds blocked from the U.S. Medical College Aptitude Test in Beirut. The two groups of students share the common language of biomedical science: each had been prepped on the Krebs cycle, DNA repair, and G proteins. Sadly, each was hunkered down at home with textbooks in hand, the issue in doubt, and death waiting around the corner. Despite their common language, they were separated by a wide gap in social history, a gap filled by hirsute terrorists who care more for their hatreds than for their children.

As Galileo and Darwin learned, the culture of science rests on fragile ground. Sweet reason is distrusted, even in the most languid of venues: look at the story of evolution in Kansas. But, let slip the dogs of war and common ground crumbles under the Katyushas. It is at times like these that we and our colleagues in the Middle East must continue to speak to each other in the collegial voice of skeptical reason. When the shooting stops, when the rockets are grounded, when blame is allotted, we ought to be sure that, as scientists, we have remained true to the charge that unites us: the search for, not the worship of, veritas. I'm pleased to note that within the past few years, the *FASEB Journal* has received and reviewed and published submissions from universities in Iraq, Sudan, Jordan, Lebanon, Israel, and the Islamic Republic of Iran. Our contributors are our brothers and sisters in the republic of science, and we stand with them as the horsemen pass by.

THE FRANCO-PRUSSIAN WAR ON CHOLERA

The Middle East was the setting for an exemplary tale of collegiality in the midst of a bitter scientific competition between two nation states.

> *Up to now twenty-two cholera victims and seventeen cholera patients have been examined in Calcutta, with the help of both the microscope and gelatin cultures. In all cases, the comma bacillus has been found. These results, taken together with those obtained in Egypt, prove that we have found the pathogen responsible for cholera.*[3]

So wrote Koch on February 4, 1884, from India, reporting on the success of the German Cholera Commission to his supervisors in Berlin. In July 1883, cholera had entered Egypt from India and the Arabian Peninsula along the recently opened Suez Canal. Quarantine facilities were set up at major ports and at way stations for pilgrims returning from Mecca, to which Muslims from India had brought the disease. The caliphate appealed to Europe for help, rightly supposing that the century-long rivalry in science between France and Germany would prompt both countries to send the best and brightest of their new microbe hunters. National honor and human lives were both at stake; national honor was clearly paramount. In 1870, the Prussians had humiliated the French on the battlefield and had succeeded in besieging Paris, the inhabitants of which were reduced to eating rats. Altruism was a secondary motive; if the disease could be halted in Egypt before it spread to the West, the fifth pandemic of Asiatic cholera might be prevented.

By mid-August 1884, the two contending teams of French and German scientists were on site in Alexandria seeking to be first to isolate the cholera agent. The French team, which was hand-picked by Pasteur himself, consisted of the internist Isador Straus, the veterinarian Edmond Nocard, and two of Pasteur's most valued assistants, Emile Roux and Louis Thuillier. Together with several Italian colleagues, they were quartered at the Hôpital Européen. The Germans were led by Koch, fresh from his triumphant discovery of the tubercle bacillus (for which he would be awarded the Nobel Prize in 1905). He was assisted by the chemist Ludwig Treskow and the bacteriologists Georg Gaffky and Bernhard Fischer; they were housed in the Greek Hospital across town from the French team. (These days, we remember Roux, Gaffky, and Nocard as names of bacteria we studied in microbiology. Their fame is shared by a dirty village stopover for pilgrims on the Suez Canal; El Tor is now simply the technical name for a bad strain of the cholera vibrio.)

The French struck paydirt soon after arrival. They found strange new microscopic structures in stained blood smears of patients with cholera. Excitement mounted, and news of the possible identification of the cholera agent spread around town. But before their preliminary finding could be confirmed, tragedy struck the French camp. Thuillier, at the age of twenty-seven the youngest of the team, succumbed within thirty-six hours to an explosive bout of cholera. As Roux reported to *Le Temps*:

> *Straus and I were obliged to hold him up to prevent his fainting. From this moment everything passed involuntarily; and, in spite of the most energetic treatment, at eight o'clock he was already moribund. . . . We employed strong frictions (rubbing of the limbs). All the French and Italian doctors were present. Iced champagne and subcutaneous injections of ether were given freely. In short, everything that could be devised was done to prevent a fatal issue.*[4]

The champagne was as useless as friction and the issue *was* fatal. At Thuillier's funeral, Robert Koch and the other German scientists showed up with two wreaths for their fallen competitor. "They are only a small token, but they are of laurel and most fitting for him, who deserves such glory," said Koch, who helped to carry the coffin. Myth and legend have arisen around this episode. For it turned out that the strange new particles discovered by the French group were nothing but fractured blood *platelets* with altered staining properties. Koch is said to have realized that the French discovery must be an artifact and suspected that the real culprit was a novel microbe that resembled a punctuation mark: *"ein Komma bazillus."*

In the legend, Koch is called out in the middle of the night by a visit from the distraught Roux. The two erstwhile rivals rush through the dark, disease-ridden town to the bedside of the dying Thuillier, whom Koch has known since the young man visited him in Berlin. Thuillier looks up at the master and weakly asks his evaluation of the new "organisms" the French have spotted in the blood. "Have we found it?" asks the moribund youth. Koch wishes his colleague to die a happy man: "Yes, you have found it."[5]

CHOLERA IN INDIA:
THE ORIENTAL CARD

In reality—as opposed to this legend from de Kruif's *Microbe Hunters*—neither the French nor German team made the conclusive discovery in Alexandria. Perhaps because of strict quarantine measures, perhaps because the epidemic wound down naturally with the end of hot weather, Thuillier's death was almost the last from cholera that season. The French went home with their samples and Koch proceeded to Calcutta. There the disease was still rampant, and it was in India that Koch isolated pure cultures of the cholera vibrio for the first time. He spelled out its local epidemiology and finally proved that it is spread via contaminated water. By rigorous bacteriological means, he showed that John Snow had been correct in 1854 when he concluded that water from the Broad Street pump spread the disease. Koch and the Europeans were entertained at clubs in the British Raj from which native Indians (called wogs for "worthy oriental gentleman") were excluded.

Once the Egyptians learned that cholera was carried by pilgrims returning from Mecca, strong sanitary measures combined with enforced quarantine along the Suez Canal stopped this route from becoming a chronic portal of entry. But the road from Mecca continued to stir the Western imagination by combining sanitary fear with religious chauvinism. In *Microbe Hunters* (1926) Paul de Kruif played the oriental card in the popular prose of his generation:

> It is thanks to these bold searchings of Robert Koch that Europe and America no longer dread the devastating raids of these puny but terrible little murderers from the Orient—and their complete elimination from the world waits only upon the civilization and sanitation of India.[6]

The microbe hunters of the nineteenth century were certain that once the causative organism of a disease was identified and its mode of spread appreciated, sanitary measures would suffice to eliminate it. And if those who carried it were the immigrant, the poor, the mad, the Gyppo (Egyptian), the Indian, or Jew—well, so be it. The needs of public health came first and the rights of "lesser breeds without the law"[7] a distant second. We get more than a whiff of this from Paul de Kruif.[8] Indeed, I'd guess that strife in today's Middle East owes not a little to Western contempt for the ignorant, unsanitary "oriental."

Ironically, a good number of my students over the last few years have been led to read de Kruif's *Arrowsmith* because someone told them that the name of the rock group "Aerosmith" was taken from a "doctor book." They've found it quaint, dated, and totally inspiring. Martin Arrowsmith, in turn, was inspired by Koch, Roux, and Thuillier. Sinclair Lewis has Martin Arrowsmith resolve that if he had to be "a small town doctor he would be such a small town doctor as Robert Koch." Unlike Aerosmith's grungy millionaires, however, Arrowsmith leaves riches not for rags, but for science; his picaresque career blends the biographies of Paul de Kruif and Sinclair Lewis's father: research in a bacteriology lab in Michigan, a small-town doctor's life in Wisconsin, work at the Rockefeller Institute, the temptations of money and the flesh. Eventually Martin and his wife travel to a Caribbean island to stop an outbreak of the pneumonic plague by means of a bacteriophage he has developed. On that plague-ridden island, Martin's Swedish colleague is killed by the epidemic they have been fighting; his death is the heroic death of Thuillier:

> *"What is it? What is it?" "I t'ink—it's got me. Some flea got me. Yes," in a shaky but extremely interested manner, "I was yoost thinking I will go and quarantine myself. I have fever all right and adenitis . . . O my God, Martin, I am so weak! Not scared . . . It hurts some, but life was a good game. And—I am a pious agnostic. Oh, Martin, give my people the phage! Save all of them—*[9]

Our new millennium has been racked by the fevers of piety, pride, and national petulance; it seems longer than a short century ago that the armies of health and reason wore the same uniform. Whatever their differences, whatever their flaws, Roux and Koch, Frenchman and German, marched together under the white ensign of sanitation. Under that banner, science went on to conquer anthrax and typhoid, diphtheria and plague. It has had less luck with unreason.

Jacques Loeb—the model for Dr. Gottlieb in *Arrowsmith*—numbered Robert Koch among the men of reason, among the *philosophes* who "dared to follow the consequences of a mechanistic science, incomplete as it then was, to the rules of human conduct, and who thereby laid the foundations of tolerance, justice and gentleness which was the hope of our civilization until it was buried under the wave of homicidal emotion which swept through the world in 1914."[10] It could be said that the dream of Enlightenment reason was interrupted in 1914 by the Guns of August to become a nightmare awaiting Judgment at Nuremberg. The Katyushas in Haifa and bombs in Beirut kept our brothers and sisters awake in the new age of Endarkenment.

16. How to Win a Nobel Prize: Thinking Inside and Outside the Box

Roger Kornberg

Arthur Kornberg

Wonder—is not precisely Knowing
And not precisely Knowing not—
—EMILY DICKINSON[1]

NOBEL ON COLUMBUS AVENUE

IN DECEMBER 2006, THE KING OF SWEDEN HONORED FIVE AMERICANS WHO for that year had won all the Nobel prizes in the natural sciences. Andrew Z. Fire of Stanford and Craig C. Mello of the University of Massachusetts shared the prize in Physiology or Medicine, Roger D. Kornberg of Stanford took the solo award in Chemistry, while the prize in Physics went to John C. Mather of NASA and George F. Smoot of UC Berkeley.

Later that year, their names, together with that of Edmund S. Phelps of Columbia, the 2006 Economics laureate, were carved into a pink granite monument at the corner of Columbus Avenue and 81st Street. The monument, dedicated to Alfred Nobel and to American Nobel laureates past and present, stands behind the American Museum of Natural History in a park named after Theodore Roosevelt, the first American laureate (Peace Prize, 1906). The names of 290 other American Nobel Prize winners have been inscribed on the sides of the monument, with ample space for more to come.

The monument in New York came to mind as I read Craig Mello's response to Adam Smith of the Nobel Foundation, who called Mello right after THAT phone call from Stockholm in October.

AS: *Well first of all many, many congratulations on being awarded the prize.*

CM: *Thank you so much.*

AS: *Where were you when you heard the news?*

CM: *I was checking my daughter's blood sugar. She has type 1 diabetes so I was actually up, one of the few, I guess, in the North Americas who was awake.*

AS: *Yes, I imagine so.*

CM: *We check her frequently and I just happened to be up, checking her blood sugar. And she had a good sugar actually, 95, which is normal.*

AS: *That's good news, yes.*

CM: *I was on my way back to bed and the phone rang.*

AS: *So two good pieces of news at once! I imagine you were thinking of other things but what was your first thought on being told?*

CM: *Well, you know, gee, that's a really hard question! You know first it's disbelief, and I don't think it sinks in quickly. I felt I was sort of too young to get it this soon and thought, if it happened, it would be a few years from now. So I wasn't ready at all.*[2]

Ready or not, forty-six-year-old Mello will join other striplings carved in granite, among them Joshua Lederberg and James Watson (aged thirty-three and thirty-four, respectively, when they received *their* telephone calls).

The Nobel monument is a recent addition to the New York scene. The slab was unveiled on a damp October day in 2003 by Mayor Michael Bloomberg, with appropriate remarks by Swedish and Norwegian dignitaries, choral music, and a stirring address by Eric Kandel (Physiology or Medicine, 2000), who paid tribute to the city's public school system of which he was a product. The audience was filled with many Nobelists, their guests, consular and civic officials, and platoons of students from high schools in the neighborhood.

As the flock of VIPs dispersed, a teenage couple made its way to the monument. Arm draped about his girlfriend's shoulder, a gangly youth pointed up at the open space under all those names: "I'm gonna to be the first black guy up there in science!"

That's why we have prizes and monuments, I thought at the time, to carve something in stone for a kid to look up to.

Reflecting on the 2006 Nobels in life science, it occured to me that science gets written in stone as a result of thinking inside the box, thinking outside the box, or simply running into pure luck. Luck, especially, since, to quote Lewis Thomas, "Chance favors the prepared grind."[3]

THINKING INSIDE THE BOX

Much of what we learn in science comes from looking in depth at what we partially understand already, something at the "not precisely knowing" level of Dickinson. But if we have a new notion of where, or how, to dig even deeper, if we end up precisely Knowing something important, the results are often astonishing. I'd call this "thinking inside the box" and the work of Roger Kornberg illustrates this sort of discovery. Joining the tools of structural biology and molecular genetics, Roger Kornberg isolated Mediator, a precisely defined complex of twenty proteins that serves as the interface between gene-specific regulatory proteins and the transcription apparatus. It's the central link in the enhancer/activator/Mediator/polII/promoter pathway of sixty proteins that has been conserved from neurospora to primate and Kornberg's pictures show precisely how that apparatus serves to spew out mRNA.[4] We already knew a good number of the moving parts, but it took Kornberg to show us how they work in concert from yeast to beast.

In his Nobel interview he explained:

RK: *The remarkable thing about the machinery is the extent to which it truly functions in the way you and I imagine or think of a machine . . .*

AS: *Do you know what speed it works at?*

RK: *It copies approximately ten DNA, or RNA (as they are called), letters per second.*

AS: *And it must achieve an extraordinary fidelity in its copying because the room for error is very small.*

RK: *It achieves very great fidelity, but beyond that it has inherent mechanisms for proof-reading and correcting errors that may be made in the process.*

AS: *Perfect machine!*[5]

Mechanical analogies seem to run in the family. Arthur Kornberg, Roger's father, received the Nobel Prize in Physiology or Medicine in 1959 (they are one of six Nobel father/son pairs and the only Americans). Kornberg *père* had discovered the enzyme that made DNA, sharing the platform in Stockholm with my mentor, Severo Ochoa, who had accomplished the synthesis of RNA. Arthur Kornberg recalled the first reaction in which DNA was put together from its constituents:

While this represented only a few micromoles of reaction, it was something. Through this tiny crack we tried to drive a wedge, and the hammer was enzyme purification.[6]

Father and son suspected that if one took the genetic machinery apart and put it together properly, one could figure out precisely how it worked.

It's what Robert Hooke predicted in 1665 when the Royal Society began its assault on not precisely Knowing:

> *By this means they find some reason to suspect that those [phenomena] confessed to be occult, are performed by the small machines of Nature.*[7]

THINKING OUTSIDE THE BOX

The opposite of the "digging-ever-deeper" approach to science is to encounter something new and puzzling, and to figure it out. The hallmarks of discoveries like these are easy to spot—in retrospect. First off, the unexpected often comes from a field far removed from the scientist's field of expertise. Secondly, it stares him in the face, like Mendel's sweet peas or Darwin's finches. Finally, it tips its hand when the seminal publication sports words like "unexpected" or "surprise" in title or abstract. Here's how RNAi was announced in the 1998 Fire, Mello paper in *Nature:*

> *To our surprise, we found that double-stranded RNA was substantially more effective at producing interference than was either strand individually.*[8]

While the role of double-strand RNA-mediated gene silencing is now written in stone, the path to its discovery was by no means straightforward: it began with petunias, led to worms, then finally on to yeast and beast. Andrew Fire remembered in his Nobel interview that:

> *Well, we were led to it pretty much by our experimental noses. The people in the plant field had done tremendous work on gene silencing and so we, sort of, were following in their footsteps in trying to sort out what was responsible for this weird silencing phenomenon in the worm.*[9]

Their experimental noses had a long trail to sniff. The history of the "Silence of the Genes," as Sen and Blau recently recorded, is, indeed, complex.[10]

Petunia flowers exhibiting sense co-suppression (RNAi) patterns of chalcone synthase silencing.[11]

In 1990, plant scientists (Carolyn Napoli, Christine Lemieux, and Richard Jorgensen) at a biotechnology company were studying enzymes that formed anthocyanin, the pigment that makes petunias purple. Testing whether chalcone synthase (CHS) was the rate-limiting enzyme in anthocyanin biosynthesis, they overexpressed chalcone synthase in petunias: "Unexpectedly the introduced gene created a block in anthocyanin biosynthesis"[12] and 42 percent of the plants became white or had chimeric purple-white patterns (see picture above). They had silenced the gene for CHS and this proved to be heritable: "progeny testing showed that the novel phenotype co-segregated with the introduced CHS gene."[13] Sure enough: "unexpectedly" was the operative word in the petunia abstract, the work it described led directly to Fire and Mello's "surprise" in worms, and to that pink granite monument on Columbus Avenue.

A LUCKY ERROR

Planning for the Nobel monument in New York began with discussions in December 2001, when the Nobel Foundation celebrated its centenary in Stockholm. Most of the living Nobel laureates in Physiology or Medicine (including a good number of Americans—Joshua Lederberg, Barry Blumberg, Alfred Gilman, Joseph Goldstein, Michael Brown, Eric Kandel, among others) showed up to revisit their grandest moment. The occasion was celebrated by lectures, symposia, concerts, and the grandest of all banquets in the splendid Town Hall of Stockholm. Guests lucky enough to have been seated at the banquet table with the senior American laureate, Tom Weller of Harvard, heard a story that described the third way to make a great discovery: make a great error and recognize your luck.

Over wines of not inconsiderable vintage, Weller reminded his dinner companions that the first time American scientists had been feted in this hall was for a dazzling cure based on luck and error. In 1934, two Harvard clinicians, George R. Minot and William Murphy, joined George C. Whipple, a Rochester pathologist, on the podium in Stockholm. Based on the wrong animal model, they had found a cure for pernicious anemia.[14]

George Richards Minot, "a Yankee of the Yankees," had been a Professor of Medicine at Harvard since 1928. He also maintained an active private practice oriented to hematology.[15] William Parry Murphy, on the other hand, was a westerner of decidedly non-patrician stock. After one year at the University of Oregon Medical School he won a scholarship to the Harvard Medical School, graduated in 1922, and in the midst of his Boston residency Minot made Murphy an offer he could not refuse.

Minot, so the story goes, was accustomed to picking young physicians of the Peter Bent Brigham Hospital as associates to run his office practice, which consisted in good part of patients with diseases of the blood and with homes on Beacon Hill. As senior resident, Murphy was next in line for this plum job, but it was customary for the young men to earn their credentials by publishing one or more papers before they started. Murphy took to the journals and found a recent report by George Whipple and Frieda Robscheit-Robbins[16] that dogs made anemic by repeated bleeding could be restored to health by feeding them huge quantities of uncooked liver. If it worked in dogs, why not in humans?

The first patients to whom Murphy fed slightly cooked liver were patients with pernicious anemia. One of these patients was an obstreperous, fretful, cantankerous old woman, whom Murphy cajoled into taking her daily ration of half a pound of liver only after a mighty contest of wills. Murphy described his everlasting "surprise": not only did her red blood cells respond to a week or so of this cumbersome regimen, but she was also relieved of her mental symptoms and soon reverted to her agreeable self. Eventually, his lucky observation of her mental improvement persuaded him that the factor in liver must work not only in the marrow but also elsewhere in the body. Iron alone could not do the trick. By May 1926, Minot and Murphy had treated each of forty-five patients with half a pound of liver a day. Soon Minot persuaded a young biochemist, E. J. Cohn, to make a liver extract rich in the anti-pernicious anemia factor. The extract briskly revved up red cell production by the marrow and cured the disease as readily as raw liver. Indeed, it remained the primary treatment for pernicious anemia until isolated and named vitamin B_{12} in 1948. It was among the first "miracle drugs" to cure a hitherto fatal disease.[17]

Ironically, Whipple's dogs had not suffered from pernicious anemia, they had iron deficiency anemia caused by repeated bleedings; and their response was due to the iron present in massive doses of liver. Whipple's error and Minot's luck led directly to the first Nobel Prize in Physiology or Medicine for work in the U.S.

But there's more to the Minot story, and it touches on that telephone call Mello received from Stockholm. Like Mello's young daughter, Minot suffered from what we would now call Type I diabetes: he was also gravely ill when he made his great discovery. A strapping six-footer, Minot had developed a severe case of diabetes; by 1922, his weight had dropped to 120 pounds.[18] As luck would have it, Charles Best and Frederick Banting in Toronto, following the digging-ever-deeper approach of John Macleod (Medicine and Physiology, 1923) had prepared the first useful batches of insulin. Minot became one of the first patients treated with the hormone by Eliot Joslin. One Nobel led to another: had insulin not been discovered, vitamin B_{12} might have been a distant dream.

Today, the names of Fire, Mello, and Kornberg have joined those of Minot and Murphy carved in stone on that pink granite slab in back of the museum in New York. More likely than not, some other kid from the neighborhood will point up at those names and start to wonder about what it takes to discover the new.

17. Homer Smith and the Lungfish: The Last Gasp of Intelligent Design

Chordates Come Ashore.
From *Kunstformen der Natur*,
Ernst Haeckel (ca. 1902)

Homer Smith, 1895–1962.
Physiologist, author of *Kamongo:
The Lungfish and the Padre* (1932)

The fishes of the early and middle Devonian found themselves forced to choose between the invading saltwater marshes and the isolated fresh-water pools which periodically contracted into stagnant swamps or hard mud flats.... The more advanced of the fishes, however, in order to survive in the stagnant waters of the continents, took to swallowing air and thus invented lungs and prepared the way for the evolution of the terrestrial vertebrates.

—HOMER SMITH, "The Evolution of the Kidney" (1942)[1]

The scientists used state-of-the-art statistical and molecular methods to unravel the evolution of the hormone aldosterone. They resurrected the ancestral receptor gene— which existed more than 450 million years ago, before the first animals with bones appeared on Earth. The experiments showed that the receptor had the capacity to be activated by aldosterone long before the hormone actually evolved.

—*Science*, 2006[2]

This weekend at the New York Auto Show, the taxicab of the future will be on display. Designers say the cab is so futuristic that the driver will be from a country that doesn't even exist yet.   —CONAN O'BRIEN[3]

EVO / DEVO AND THE CONQUEST OF LAND

CREATIONISTS, FANS OF "INTELLIGENT DESIGN," AND OTHER FLAT-EARTH zanies have been playing defense since December 2005, when Judge Jones of Dover, Pennsylvania, ruled that intelligent design (ID) had no place in science classrooms. After Dover, bills designed to subvert the teaching of evolution failed to pass even in such conservative states as Kansas, Utah, and South Carolina.

Meanwhile, newer studies in evolutionary and developmental molecular biology (Evo/Devo) have left the favorite notions of intelligent design in Devonian mud. At Dover, the ID apologists advanced the concept of "irreducible complexity," i.e., some systems are so complex that they had to be designed in one flash by one intelligence—guess Who?[4] It turns out that molecular genetics can reach into the Devonian seas of 400–500 million years ago to explain "Aldosterone and the Conquest of Land."[5] Evo/Devo has puzzled out a system as "irreducibly" complex as the human kidney and its hormonal regulation. Random mutations, over time, in marine creatures adapted a system that responded to stress in the sea (via cortisone) to one that retained salt and water when lungfish crawled ashore (via aldosterone). Genes for the cortisol receptor anteceded those of the aldosterone receptor *and* the p450 enzyme required to make the salt-and-water hormone. The receptor was there before the ligand existed.[6] As Conan O'Brien quipped, the New York taxi was there before the country the driver came from was invented.

ONTOGENY RECAPITULATES PHYLOGENY (SORT OF)

But don't count the faithful out. The pulpits have sharpened their attacks on Darwin and company, while the erstwhile presidential candidate Sam Brownback blamed human embryonic stem-cell research for "the destruction of young human lives."[7] Unlike the intelligent design folks, creationists like Brownback, et al. take particular offense at the notion of Evo/Devo, the academic discipline based on interactions between evolution and developmental biology. They take straight aim at Ernst Haeckel, whom they hold accountable for having launched Evo/Devo in the first place. Creationists join their belief that Darwin contradicts scripture to the conviction that Haeckel is the devil incarnate; wasn't he the German who claimed that little human embryos look like fish or larval salamanders? *Creation* magazine points the finger:

Most people have heard of or been taught the idea that the human embryo goes through (or recapitulates) various evolutionary stages, such as having gills like a fish, a tail like a monkey, etc., during the first few months that it develops in the womb. The idea . . . has not only been presented to generations of biology/medical students as fact, but has also been used for many years to persuasively justify abortion. Abortionists claimed that the unborn child being killed was still in the fish stage or the monkey stage, and had not yet become a human being. This idea (called embryonic recapitulation) was vigorously expounded by Ernst Haeckel from the late 1860s to promote Darwin's theory of evolution in Germany, even though Haeckel did not have evidence to support his views.[8]

Not quite. First of all, the early human embryo *does* look like a little fish (almost) with a tail and branchial clefts under the chin, just where gills develop in lower forms (more or less). Haeckel himself argued that recapitulation doesn't imply identity.[9] Second, Haeckel expanded Darwin's description of natural selection to include the rest of nature. It's no accident that Joe Thornton, who made the aldosterone discovery, works at Oregon's Ecology and Evolutionary Biology Center. Haeckel coined the word "ecology" in 1866 to describe the study (*logos*) of the house (*oikos*) of nature: Darwinian selection takes place in a changing environment. Finally, Haeckel's phylogenetic drawings have been reexamined and found to be more illustrative than evidentiary. Nevertheless, their heuristic value is now generally accepted: "Haeckel recognized the evolutionary diversity in early embryonic stages, in line with modern thinking. Haeckel's much-criticized embryo drawings are important as phylogenetic hypotheses, teaching aids, and evidence for evolution. . . . Despite his obvious flaws, Haeckel can be seen as the father of a sequence-based phylogenetic embryology."[10]

To the legions of the faithful, however, Haeckel remains bad news. From a website aptly named "Apologetics Press" comes the old canard: "Dr. Haeckel was an accomplished artist, as well as an anatomist. He used his art talent to falsify some of the drawings that accompanied his research articles on animal and human embryos, in order to make it appear as if embryonic recapitulation were true—when, in fact, it was not."[11] Scores of high-tech websites and amateur blogs point the fickle finger of shame on Haeckel. Folks who insist that the fossil record "proved the Bible and God right and evolution wrong"[12] and who cite junk science to explain that "the only way we know for a DNA to be altered is through a meaningful intervention from an outside source of intelligence,"[13] also complain that schools and churches have been subverted by "academics kowtowing to every pronouncement made by Darwin and his God-hating successors"—like Haeckel.[14] The Dells are alive with the sound of malice.

FROM FISH TO PHILOSOPHER

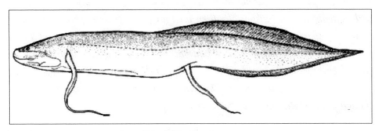

The African lungfish, *Protopterus aethiopicus aethiopicus*

Haeckel's notion of the phylogenetic law received early approval from Darwin. In his *The Descent of Man* (1871), Darwin chimed in with:

> *He who wishes to see what ingenuity and knowledge can effect, can consult Dr. Haeckel's work. . . . As the class of fishes is the most lowly organized and appeared before the others we may conclude that all the members of the vertebrate kingdom are derived from some fishlike animal, for they [all] have much in common, especially in their embryonic stage.*[15]

The clearest example of "having much in common in their embryonic stage," i.e., phylogeny recapitulating ontogeny, is that of the human kidney. Homer Smith's classic *The Evolution of the Kidney* (1942) puts it neatly: "In the ontogenic development of the human embryo, the glomerulus is not brought into conjunction and connected with the tubule of the metanephros until some time after the tubule has been formed . . . this interval between the development of the tubule and the glomerulus is an ontogenic recapitulation of the phylogenetic interval which separated their evolution."[16]

Smith, Professor of Physiology at NYU School of Medicine, was as broadly cultivated, as fine an artist—albeit in prose—and as influential in his own field as Haeckel; he was recognized as dean of renal physiology at mid-twentieth century.[17] His contributions included the introduction of inulin to measure kidney function, the use of dyes as such as p-aminohippurate to measure renal blood flow, and development of the concept of Tm, the maximal rate at which a substance is taken up by the kidney from blood or renal tubules. Together with Herbert Chasis and William Goldring, he showed that much of high blood pressure in humans is due to peripheral constriction of small blood vessels—an observation that made possible today's treatments for high blood pressure.

Smith was also a skeptical thinker, whose many essays, lectures, and books echoed the humanist tradition of the first chair of Physiology at NYU, John William Draper (*History of the Conflict Between Religion and Science,* 1874). In praising the joys of here and now over the promise of pie in the sky, he'd repeat Isak Dinesen's jibe, "What is man, when you come to think upon him, but a minutely set, ingenious machine for turning, with infinite artfulness, the red wine of Shiraz into urine?"[18] He had a higher view still of renal function: "Superficially it might be said that the function of the kidneys is to make urine; but in a considered view one can say that the kidneys make the stuff of philosophy itself."[19]

THE LUNGFISH AND THE PADRE

By the age of thirty-seven, Homer Smith had become a national celebrity for his best-selling book *Kamongo, or The Lungfish and the Padre* (1932). An eloquent exercise in popular Darwinism, the book remained in print for twenty years. Smith had traveled to the papyrus marshes of eastern Lake Victoria, to bring back living samples of the African lungfish (kamongo to the natives) in order to apply his inulin methods to a species in which glomerulus and tubules had not quite meshed. Lungfish have twin lungs and can breathe in both air and water, but must have air available or drown. Its lungs permit it to survive in times of drought or dust, when the lungfish encysts itself deep into soft mud.

> *When water drains away, the fish can at last breathe without moving, and it curls up with its tail across the top of its head, covering the eyes. Its body is coated with a slimy mucus secreted by the skin, and as this mucus dries it hardens into a brown, parchment-like, waterproof cocoon that envelops the body closely, extending into all exposed crevices. The only opening is a short funnel where the cocoon extends between the lips and teeth, and through which the fish breathes.*[20]

Lungfish can live in their muddy chamber for up to four years in dry spells, but by and large the rains return within half a year. They drown in neat water, and dry out in arid air. Perhaps that's why their remnants occupy such a tiny niche in our vast ecosystem.

Kamongo, the book, is cast as a dialogue between an American scientist, Joel (Smith's alter ego), and an Anglican missionary priest. The setting is the voyage home from East Africa north to the Suez Canal aboard the "S. S. Dumbea" [*sic*]. Joel and the priest talk of cosmic stuff and evolution. Joel tells the padre the story of the lungfish, a dead end in evolution; and the two have very different views of what the story means to mankind. Joel has concluded that the notion of any overriding universal intelligence is an illusion. Man's brain, argues Joel, may be like kamongo's lung, both ingenious developments but neither leading to ulterior ends. Joel finishes his argument by calling life "an eddy in the Second Law of Thermodynamics." The priest, sympathetic throughout as a character, remains unconvinced and responds with an appeal to the ineffable. "Miracles happen," he sighs. The two travelers part, neither understanding the other.

Kamongo was written only a few years after the Scopes trial, and the teaching of Darwinian evolution was still illegal in several states. Evolution and Creationism were in the air and Smith had touched a public nerve. The book was picked by the Book-of-the-Month Club in April 1932, the same year that Faulkner's *Light in August* and James T. Farrell's *Studs Lonigan* shared the honor.

EVO/DEVO AND EINSTEIN

In the public sphere not too much has changed since *Kamongo*; we've now had Dover, followed by Richard Dawkins on the bestseller list. But in the realm of Evo/Devo, more has happened. Jean Joss, of McQuarrie University in Australia, has suggested that Devonian lungfish probably included metamorphosis in their life cycle.[21] They have an enormous genome, 132 pg/nuclear DNA vs 3.4 for humans or other mammals;[22] that brackets lungfish with the large genomes of larval salamanders. (See Haeckel's illustration, above.) In this modern tree of life, the larval forms of Devonian lungfish became the direct ancestors of tetrapods, and, needless to say, our ancestors as well. And in utero, our twin, fetal lungs begin as small paired pouches under the chin. Devo recapitulates Evo, as Haeckel predicted.

Finally, on the intelligent design front, I'd go along with Albert Einstein in his "Introduction" to Homer Smith's last book:

> *Homer Smith's Man and His Gods (1952) is a broadly conceived attempt to portray man's fear-induced animistic and mythic ideas with all . . . the boundless suffering which, in its end results, this mythic thought has brought upon man. This is a biologist speaking, whose scientific training has disciplined him in a grim objectivity rarely found in the pure historian. His historical picture closes with the end of the nineteenth century, and with good reason. By that time it seemed that the influence of these mythic, authoritatively anchored forces which can be denoted as religious, had been reduced to a tolerable level in spite of all the persisting inertia and hypocrisy.*[23]

If we're lucky, the influence of those mythic forces can be reduced to a tolerable level once more.

18. DDT Is Back: Let Us Spray!

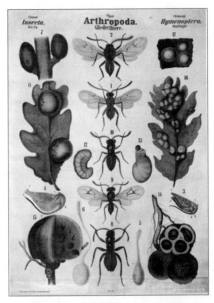

Arthropoda (1880), Rudolph Leuckart

No book, including The Protocols of the Elders of Zion, has killed more people than Silent Spring. It has done far more damage than DDT ever did. We hope its pernicious influence is finally at an end.
— *The Sunday Telegraph*, Editorial, September 17, 2006[1]

We must take a position based on the science and the data. One of the best tools we have against malaria is indoor residual house spraying. Of the dozen insecticides WHO has approved as safe for house spraying, the most effective is DDT.
—Dr. Arata Kochi, WHO[2]

A PERNICIOUS INFLUENCE

Yes, DDT is back. In September 2006, the World Health Organization held a press conference to promote the widespread use of DDT in Africa. Dr. Arata Kochi, head of the WHO malaria program, assured the world that DDT is not only the most effective insecticide against malaria but also poses no health risk when properly used. "Expanding its use is essential to reviving the flagging international campaign to control the disease," he said.[3]

The WHO announcement marked the end of a tough campaign by public-health officials and malaria experts who had argued for years that DDT was a necessary public-health weapon in poor tropical countries. The arguments were formulated in 1996 by Amir Attaran of Harvard University's Center for International Development.[4] After hundreds of physicians the world over had signed a petition urging resumption of DDT spraying, Attaran concluded that indoor residual house-spraying with DDT was "an inexpensive, highly effective practice against malaria . . . the quantities involved are minimal (2g/m²) unlike agricultural uses which inject tons of DDT into the outdoors." For the amount of DDT used on a cotton field (levels referred to by Rachel Carson in *Silent Spring*) all the high risk residents of a small country can be protected from malaria."[5] And DDT is not only effective against the mosquito vectors of malaria, it is equally robust at stopping other arthropod-borne killers such as dengue, yellow fever, sleeping sickness, and typhus. It has played a major role in twentieth-century history.

DDT was first introduced to the world by Swiss chemist Paul Hermann Müller (1899–1965) of Geigy AG. Müller passed its secret to the Allies toward the end of World War II, and wherever they employed the pesticide, arthropod-borne diseases were eradicated. Lacking the pesticide, German soldiers died of typhus by the thousands, their prisoners in flea-ridden concentration camps by the hundreds of thousands: typhus was the second leading cause of death in the camps.[6]

Müller was awarded the Nobel Prize in Physiology or Medicine in 1948 for his "discovery of the high efficiency of DDT as a contact poison against several arthropods." The full citation gives him proper credit: "DDT has been used in large quantities in the evacuation of concentration camps, of prisoners and deportees. Without any doubt, the material has already preserved the life and health of hundreds of thousands."[7] Shortly thereafter, DDT became available worldwide and was successfully employed against other arthropod-borne diseases, most notably malaria. DDT helped to eradicate malaria from the developed regions of the world (the United States, Europe) and to lower its incidence by over 99 percent in others (Sri Lanka, India). In South Africa's KwaZulu-Natal province, malaria epidemics before DDT killed more than 22,000. By 1973, after DDT, South Africa recorded only 331 malaria cases in the entire country; in 1977, only a single death was reported.[8] Malaria was on its way out in sub-Saharan Africa.

SILENT SPRING

And then came the political fallout from Rachel Carson's 1962 bestseller, *Silent Spring,* a passionate plea for a clean earth.[9] Rachel Carson (1907–1964), a fine naturalist and longtime marine biologist at Woods Hole, Massachusetts, called popular attention to DDT's effects on the eggs of raptor birds and seashore life. She won public acclaim for her splendid prose and even greater public sympathy for her poignant situation. Carson's *cri du coeur* for an end to chemical pollutants was written as she was dying of breast cancer. Chemicals, she wrote,

> *have the power to kill every insect, the good and the bad, to still the song of birds and the leaping of fish in the streams, to coat the leaves with a deadly film, and to linger on in the soil. . . . Can anyone believe it is possible to lay down such a barrage of poisons on the surface of the earth without making it unfit for all life? They should not be called "insecticides" but "biocides."*[9]

She was correct, of course, and her indictment of the agricultural/industrial complex was the beginning of modern ecological wisdom. By the mid-70s the WHO dropped its DDT program worldwide, having been convinced by Carson's followers that indiscriminate agricultural use of DDT, especially in developed countries, had polluted soil and stream and ocean. By then, millions of tons of DDT had been spread over farmland and forest, suburb and seashore. A generation after DDT was banned, legitimate questions remain as to whether low levels of environmental DDT can stunt the fertility, growth, and development of creatures great and small.[10]

DDT IS OUT

Sadly, by the 70s, the baby of disease control had been thrown out with the (polluted) groundwater. DDT was banned in most corners of the earth, with its unintended consequences falling on the poorest countries. Thanks to DDT, countries such as Zanzibar had reduced the percentage of their populations infected with malaria from 70 percent in 1958 to under 5 percent in 1964. When the DDT spraying was halted, the malaria rate rose back to over 50 percent by 1984.[11] Swaziland, which did not halt DDT spraying, maintained its malaria infection rates between 2 to 4 percent, while just forty miles away, South Africa, which had banned DDT in the 80s, watched malaria infection rates rise to 40 percent. In the mid-90s South Africa and a few other African countries resumed spraying with heartening results: wherever DDT has been used in a campaign of indoor residual house-spraying, malarial rates fell.[12] And yet, in response to the concerns of environmental activists, mainly in developed lands, malaria continues to exert its toll. Today, there are close to 3 billion people at risk for the disease, with 500 million cases each year, causing between 1 and 3 million deaths.[13]

ARTHROPODS WITHOUT BORDERS

But that's just the toll of malaria. Mosquito-borne dengue fever is estimated to afflict 50–100 million people each year and to kill 50,000 worldwide in Africa, South and Central America, and Southeast Asia.[14] The WHO estimates that fly-borne East and West African trypanosomiasis is a risk for 50 million, with 20,000 newly reported patients a year, most of whom will die. The South American form of trypanosomiasis, Chagas disease, threatens 100 million, of whom 4.5 million will die from its chronic insults. Of the adult populations at risk for Chagas disease in South and Central America, 10 percent may die.[15] More yet: Each week there are newer outbreaks of diseases borne by mosquitoes, ticks, fleas, and lice. Two generations after Müller, epidemic typhus is making a comeback. A single outbreak of "jail fever" in Burundi sparked an extensive epidemic of louse-borne typhus in the refugee camps of Rwanda, Burundi, and Zaire—countries racked by ongoing civil war and genocide.[16] The lice are also biting in the highlands of Algeria.[17]

Lice require dirty humans, bad weather, and crowding—as in tents and barracks. That's why typhus was the stuff of war and tragedy. Even before rickettsiae were identified, it was appreciated that epidemiology and public health might be better weapons against typhus than drugs or antisepsis. In 1848, a young Rudolf Virchow (1821–1903), the future father of cellular pathology and future leader of the Social Democrats in the Berlin parliament, was sent to Silesia to study an intractable typhus epidemic in Upper Silesia, at the eastern border of Germany and Poland. In this backward enclave, feudal landlords of large estates ruled a peasantry in dirty huts. Virchow reported to Berlin that "the Upper Silesian in general does not wash himself at all, but leaves it to celestial providence to free his body occasionally by a heavy shower of rain from the crusts of dirt accumulated on it. Vermin of all kinds, especially lice, are permanent guests on his body."[18] Virchow concluded that what the region needed was not more doctors; it required social medicine for a social disease. Typhus could only be eradicated, Virchow argued, by means of political reform: full employment, higher wages, the establishment of agricultural cooperatives, universal education, and the disestablishment of the Catholic Church.[19] And slowly but surely, from Bismarck's universal health insurance to the Weimar Republic's secular reforms, typhus abated.

Less than one hundred years later, The German Endarkenment had undone social democracy in its first home and, predictably, typhus returned. Its most poignant episode was played out between March 31 and April 15, 1945, at Bergen-Belsen. Anne Frank was one of eight Dutch Jews who had been in hiding for two years and thirty days when they were discovered and arrested by the Nazis and deported from Amsterdam to Auschwitz-Birkenau. As the Russian armies moved west, the Germans forced thousands of men, women, and children across war-torn Europe to Bergen-Belsen, a desolate concentration camp north of Hannover. Among them were Anne and her elder sister Margot, who died of typhus at Bergen-Belsen, along with approximately 50,000 of their fellow prisoners.[20] The last woman to see the sisters alive described the hut in which they were quartered:

> *The dead ones were always carried out and put in front of the barrack . . . on one of those trips to the latrine or barrow I passed the corpses of the Frank sisters, one or both, I know not. And those stacks of corpses were cleared away. There was a large pit dug in the camp and they must have been thrown into it.* (Es wurde eine große Grubegegraben, da wurden sie hineingeschmissen, so kann man es wohl sagen.)[21]

On April 15, 1945, when Bergen-Belsen was liberated by a combined British-Canadian unit, the camp had been without food or water for three to five days. Although the camp commandant, Josef Kramer, protested that there was no way to pipe water into the camp, the Allied unit quickly constructed a makeshift piping system from a nearby river to supplement the army's water carts. Clean water flowed and DDT followed.

The stories of Anne Frank, of Burundi and Zaire, confirm that Virchow was right: typhus—not to speak of malaria—is a social disease bred in cruelty and based on misery. Control the vector (human or arthropod) and one controls the disease. Were plucky Rachel Carson alive, I'd bet that she would be among the first to support a WHO campaign to spray those camps in Rwanda, Burundi, and Darfur. Even with DDT.

19. Academic Boycotts and the Royal Society

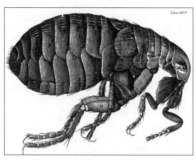

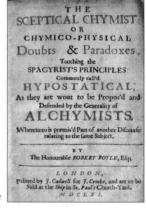

Above, Robert Hooke, "The Flea," in *Micrographia* (1661 ed.)

Right, Robert Boyle's *Sceptical Chymist*

The University and College Union [UK], representing more than 120,000 college-level educators, voted May 30 to pass a controversial resolution calling for a boycott of Israeli academics and universities. —The Economist, June 16, 2007[1]

Martin Rees, the President of the Royal Society, re-affirmed the commitment of the Royal Society, the UK national academy of science, to a statement in opposition to academic boycotts. "Moratoria on scientific exchanges based on nationality, race, sex, language, religion, opinion and similar factors thwart [our] goals. —E. B. DAVIES, FRS, July 4, 2007[2]

The other humane studies I apply myself to are natural philosophy, the mechanics and husbandry, according to the principles of our new philosophical college . . . an Invisible College. —ROBERT BOYLE, FRS, 1646[3]

Dr. South [Canon of Christchurch] made a long oration of satirical invectives against Cromwell, fanaticks and the new philosophy. . . . "They can admire nothing except fleas, flies and themselves." —JOHN WALLIS, FRS. to Robert Boyle, FRS, Oxford, 1669[4]

BOYCOTT US

IN THE SUMMER OF 2007, LAPTOPS ON BENCHES AND BEACHES WERE AWASH in protests against an academic boycott of Israel launched by a union of British educators. The boycott resolution, passed by the University and College Union (UCU) of Britain, calls for union members to cut their contacts with Israeli academics and universities, and to stop EU funding of Israeli scientific research. The motion would also require UCU members not to submit articles to Israeli journals.[5] The UCU action was taken in response to an appeal by a Palestinian union of academics which urged their colleagues in the UK "to comprehensively and consistently boycott all Israeli academic and cultural institutions and to refrain from participation in any form of academic and cultural cooperation, collaboration or joint projects with Israeli institutions . . ." as protest against "military occupation and colonization."[6]

The boycott motion was passed at the inaugural meeting of a new union formed by a merger of the Association of University Teachers (AUT) with the National Association of Teachers in Further and Higher Education (NATFHE). Although the group now represents 100,000 members, only 257 members took part in the ballot, with 158 in favor and 99 against. The resolution petered out over the next year, but its impetus remains alive.

I'd argue that the resolution set a terrible precedent because discourse without borders makes science happen. The *FASEB Journal* has, over the years, received submissions, reviews, comments, and letters from Egypt, Lebanon, Iran, and the United Arab Emirates, as well as from Israel and the UK. This must continue. As experimental scientists we are not only citizens of one or another country, members of one or another sect, but we are also members of what Robert Boyle called the "invisible college" of natural philosophy.[7] The Royal Society, the original invisible college, has rightly opposed any such boycott because it would "deny our colleagues their rights to freedom of opinion and expression; interfere with their ability to exercise their bona fide academic freedoms; inhibit the free circulation of scientists and scientific ideas; and impose unjust punishment."[8]

Other colleges, visible and invisible, have also protested the boycott. In the UK, the resolution was denounced by the Academy of Medical Sciences and from Universities UK, a group representing vice chancellors of twenty research universities. In New York, the Wiesel Foundation chimed in with a strong statement in support of academic freedom signed by fifty-one Nobel laureates, including Mikhail Gorbachev, the Dalai Lama, and former South African president F. W. de Klerk.[9] The American Physiological Society voted to reaffirm their opposition to "all discrimination on the basis of such factors as ethnic origin, religion, citizenship, language, political stance, gender, sex or age."[10] Under an ad hoc banner, more than 10,000 scholars, including thirty-two Nobel laureates and fifty-three university and foundation presidents, expressed outrage at the UCU, declaring, "We Will Regard Ourselves as Israeli Academics and Decline to Participate in any Activity from which Israeli Academics are Excluded."[11] Full page ads in the *New York Times,* signed by scores of university presidents, featured a similar challenge by Lee Bollinger of Columbia: "Boycott us, then, for we gladly stand together with our many colleagues . . . against such intellectually shoddy and politically biased attempts to hijack the central mission of higher education."[12]

Writing as an editor, I'd stand with Lee Bollinger: "Boycott us, then, if you boycott any journal of science, in any corner of the world."

BOYCOTT, SCHMOYCOTT

Unfortunately, the UCU-proposed boycott carried with it more than a hint of age-old bigotry, since no similar boycotts have been advocated against states other than Israel. States that—in the words of the Palestinian appeal to which the UCU responded—could be accused of "military occupation and colonization."[13] Were there no British troops in Basra?

Political scientist Geoffrey Alderman of Buckingham University has analyzed why Israel has been singled out for boycott by the new British union. Alderman reminds non-Brits that "a very substantial number of members of the UCU are not, and never have been, academics." Rather, NATFHE functioned as an old-fashioned trade union of educational employees, sharing a traditional Marxist antipathy to Jewish nationalism and mandarin scholarship. "Add to this a goodly measure of anti-colonialism, and a stubborn determination to view Israel merely as a colonial outpost of Anglo-America, and you have a fertile ground in which the boycotters were able to plant their seeds."[14]

ACADEMIC FREEDOM AND "EXPERIMENTALL LEARNING"

A professional historian of science might remind his colleagues in the UCU that the seeds of modern academic freedom were sown in the same soil as that of Marxist theory. The Prussian constitution of 1850, which was a direct consequence of the European revolutions of 1848, established *Freiheit der Wissenschaft* (freedom of science) as a university standard. It became the third academic freedom, so to speak, alongside the older principles of *Lehrfreiheit und Lernfreiheit* (freedom of teaching and learning). And as German *Wissenschaft* grew to be the envy of the world in the course of the nineteenth century, its tradition of *Freiheit* protected dispute and dissent in every grove of academe.[15] Hermann von Helmholtz quarreled with Edward Hering over human perception; Paul Ehrlich and Robert Koch disagreed with each other (and with Julius Cohnheim) on immunity; while Marx's pal Friedrich Engels,[16] Rudolf Virchow and Ernst Haeckel[17] held very different views of natural selection. There was academic *Freiheit* aplenty for Darwinist and Lamarckian, nobleman and communist, Jew and Gentile. Alas, less than a century later, it all ended with national socialism . . . and then there was Israel.

That same historian might adduce a lesson for the British educators closer to home. The Royal Society of London was founded in an age of sectarian violence, political chaos, and brutal regicide: there was that small matter of King Charles's decapitation. There was also enough "military occupation and colonization" in the British Isles to last for three centuries. There was also that small matter of Cromwell sacking Ireland. But, thanks to the efforts of men from each side of the seventeenth-century civil wars, "a Colledge for the Promoting of Physico-Mathematicall Experimentall Learning"[18] finally emerged. John Wallis left a first-person account of how the invisible college became the Royal Society:

> *About the year 1645, while I lived in London (at a time when, by our civil wars, academical studies were much interrupted in both our Universities) . . . I had the opportunity of being acquainted with divers worthy persons, inquisitive of natural philosophy, and other parts of human learning; and particularly of what hath been called the New Philosophy or Experimental Philosophy. We did by agreements, divers of us, meet weekly in London . . . About the year 1648-49, some of our company being removed to Oxford (first Dr Wilkins, then I, and soon after Dr Goddard) our company divided . . .*

The London group continued to meet at Gresham College until the year 1658, when the they had to disband in fear of their lives as soldiers took over their meeting rooms and London underwent a period of terror. In February 1660, Monk's army entered London and restored order. King Charles returned to London at the end of May 1660 and the meetings at Gresham College resumed.[19]

Following the restoration of Charles II, the first meeting of the Royal Society was held at Gresham College on November 28, 1660. Christopher Wren, Gresham Professor of Astronomy, delivered a lecture on the pendulum to an apostolic audience of twelve. These included six roundheads (parliamentarians), four cavaliers (royalists), and two with their positions on regicide undefined.[20] Would that Sunnis, Shias, and Kurds could assemble such a college in post–civil war Iraq!

LORDS OF THE FLIES

The new philosophers chased from Oxford by the Restoration were mainly of roundhead allegiance. Royalist clerics like Robert South of Christchurch were concerned that science had entered Oxford behind Cromwell's parliamentary armies—not entirely wrong, since members of Boyle's invisible college included the first-secretary-to-be of the Royal Society, John

Wilkins, Cromwell's brother-in-law; Jonathan Goddard, who was a physician to Cromwell's armies in Ireland and Scotland; William Petty, who was the surveyor for these armies; and John Wallis (see above), who was the cryptographer for Parliament.[21]

But the invisible college had room for cavalier as well as roundhead, scientist as well as amateur. The first president of the Royal Society was Viscount William Brouncker, a royalist physician and mathematician who went on to formulate the generalized, continued fraction of π^{mth}. Brouncker's father had bought himself into the Irish peerage and, according to diarist Samuel Pepys, another FRS, "gave 1200 pounds to be made an Irish lord, and swore the same day that he had not 12 pence left to pay for his dinner."[22] In 1662, Charles II appointed William Brouncker as Keeper of the Great Seal. Brouncker proceeded not only to keep the seal but also to bestow it on the Colledge for Experimentall Learning—and the Royal Society was born.

Another peer, the Honorable Robert Boyle, was rightly called "the father of chemistry and the son of the Earl of Cork" and gave the new science its bona fides by virtue of his rank. Boyle is remembered by every high school student, if not by every pulmonary physiologist, as the contributor of V (for volume) to the law of perfect gases $(PV = n\,RT)$. But perhaps Boyle's major achievement was to distinguish the facts of chemistry from the opinions of alchemy. It was Boyle who finally dispensed with the old notions of "elemental" earth, air, fire, and water, announcing in *The Sceptical Chymist* (see above) the new Restoration definition of an element:

> *Certain primitive and simple, or perfectly unmingled bodies; which, not being made of any other bodies, or of one another, are the ingredients of which all [other] bodies are immediately compounded, and into which they are ultimately resolved.*[23]

Robert Hooke, the butt of Robert South's taunt that the royals cared for nothing but "fleas, flies, and themselves," did not come from noble stock. Indeed, he may have been the first professional scientist of the modern era, starting his career as Boyle's lab assistant. Eventually he was employed by the Royal Society as its "curator of experiments," i.e., he arranged bench-top demonstrations at its meetings. Hooke not only made microscopy practical, but in his atlas of magnificent engravings *Micrographia* (1663), first suggested that living matter, as in cork or sponge, was composed of smaller subunits that resembled the cells of monks:

> *I could exceedingly plainly perceive it to be all perforated and porous, much like a Honey-comb, but that the pores of it were not regular. . . . these pores, or cells were indeed the first microscopical pores I ever saw, and perhaps, that were ever seen.*[24]

His most famous image was that of the flea, which he described as "adorn'd with a curiously polish'd suite of sable Armour, neatly jointed."[25] Hooke, as mere commoner, became a prime target for such enemies of the new science as the Canon of Christchurch and the Bishop of Ely, who could not risk offending the likes of Viscount Brouncker or Boyle. Hooke and his insects also became the stuff of popular satire. Thomas Shadwell, probably the least gifted poet laureate of England, described Hooke in pejoratives that sound familiar to those who pursue basic science:

> *A sot that has spent two thousand pounds in microscopes to find out the nature of eels in vinegar, mites in a cheese, and the blue of plums which he has subtly found out to be living creatures. . . . One who has broken his brains about the nature of maggots, who has studied these twenty years to find out the several sorts of spiders, and never cares for understanding mankind.* [26]

SCIENCE WITHOUT BORDERS

History suggests that it would have been better for the British to have spent a few more pounds on microscopes and studies on the "nature of maggots." For when the great bubonic plague struck London in the summer of 1665, it claimed seventy thousand victims, of a population of almost half a million. Since the bubonic plague is caused by transmission of *Yersinia pestis* from infected rats to humans by those "neatly jointed" fleas, perhaps more attention should have been paid to those creatures in "sable armour" than to vestry robes or Restoration comedies.

Under Charles II, cavaliers and roundheads joined in common cause to fight the Dutch for dominion of the sea, and—one may add—to pluck New York from the Dutch (1664). But national disputes did not halt the discourse of science. Ironically, when news of the microbial world reached the Royal Society, it was from the land of Britain's bitterest adversary and from a man who could neither write nor read English. Despite national rivalries between their two countries, despite wars of "military occupation and colonization" over four continents, England's Royal Society received and published a crucial communication from Holland's Antonie van Leeuwenhoek, a lens-grinder and tradesman of Delft.[27] In April 9, 1676, in order to discover "the cause of the hotness or power whereby pepper affects the tongue," Leeuwenhoek had ground pepper seeds into water. Three weeks later, he discovered an incredible number of very little "animalcules": it was the first sighting of the microbial world. A friendly Dutch physician translated his observations and communicated them to the Royal Society. After several tries, Robert Hooke, as Curator of Experiments, confirmed Leeuwenhoek's observations, the *Transactions of the Royal Society* recorded the event, and in time the low-born lens-grinder from Holland was elected a Fellow of the Royal Society in 1680. International discourse and the world of experiment had won out over the trivial divisions of class, credentials, or country.

20. Teach Evolution, Learn Science: John William Draper and the "Bone Bill"

FASEB motto 2007

The experience of sequencing the human genome, and uncovering this most remarkable of all texts, was both a stunning scientific achievement and an occasion of worship.
—FRANCIS S. COLLINS, *The Language of God*[1]

Our discovery put an end to a debate as old as the human species: Does life have some magical, mystical essence, or is it, like any chemical reaction carried out in a science class, the product of normal physical and chemical processes? Is there something divine at the heart of a cell that brings it to life? The double helix answered that question with a definitive No. —JAMES D. WATSON, *DNA: The Secret of Life*[2]

AHEAD OF TURKEY, BEHIND IRAN

IT'S CLEAR FROM THE STATEMENTS OF FRANCIS COLLINS AND JIM WATSON that the religious beliefs of any individual scientist have no bearing on the validity or significance of the work. With Watson as its founder and Collins its executor, the public Human Genome Project was fueled by the energy of agnostic and believer alike. But what about the larger society in which these attitudes compete: Do national choices among one or another systems of belief influence the course of science and how it is taught?

Nothing illustrates the influence of religious belief more than our nation's recurring battle over the teaching of Darwinian evolution. Small, but significant, victories in Pennsylvania and Kansas have been countered by proposals in several other states, Ohio most recently, to "teach the controversy," as if intelligent design were a proposition that could explain microbial resistance to penicillin. Sad to say, while it's hard to find a modern biologist who does not accept Theodosius Dobzhansky's aphorism that "nothing in biology makes sense except in the light of evolution,"[3] the bulk of Americans have trouble making sense of Darwin.

A recent study published in *Science*[4] documents that fully one-third of American adults believe that evolution is "absolutely false" while only 14 percent of adults acknowledge that evolution is "definitely true." In Iceland, Denmark, Sweden, and France over 80 percent of adults had no trouble accepting the facts of evolution, nor did 78 percent of Japanese. Indeed, of thirty-three countries surveyed as to their acceptance of evolution, the United States ended up as thirty-second on the list. Turkey finished dead last, while Cyprus beat us by a whisker. Perhaps in response to the "intelligent design" movement, the percentage of U.S. adults accepting evolution has actually *declined* over the last twenty years. Supporting this notion, Miller, et al., found that "the total effect of fundamentalist religious beliefs on attitudes toward evolution (using a standardized metric) was nearly twice as much in the United States as in the nine European countries." They concluded that "individuals who hold a strong belief in a personal God and who pray frequently were significantly less likely to view evolution as probably or definitely true than adults with less conservative religious views."[5]

Thus, while Francis Collins may have resolved the conflict between scripture and evolution in his own mind, fundamentalism has separated Americans not only from Europe and Japan but from other parts of the globe. A bizarre editorial in *Nature,* effulgent with praise for the leader of Iran, Mahmoud Ahmadinejad, contrasted his pro-science policies with those of Christian (read U.S.) attitudes towards evolution and stem-cell research:

> One practical advantage for science in Muslim countries is the lack of direct interference of religious doctrine, such as exists in many Christian countries. There has never, for example, been a debate about Darwinian [sic] evolution, and human embryonic stem-cell research is constrained by humanistic rather than religious ethics.[6]

A HOT-BUTTON ISSUE

The split between our country and the rest of the world was brought home to me at Woods Hole last summer. Hundreds of FASEB buttons urging citizens to TEACH EVOLUTION, LEARN SCIENCE were snapped up eagerly by packs of students, post-docs, and lab people from all over the United

States. But while folks from Maine to Mississippi knew what the fuss was about, many of the European scientists who work at Woods Hole each summer simply shook their heads. "Only in America . . ." mourned an Italian microscopist; "Poor America . . ." sighed a young German biochemist. They were assured that nowhere else in the world is the Scopes "monkey trial" replayed daily in the courts. And nowhere else in the world is a leader of modern science likely to argue that:

> *You've got to accept who Christ was and what He said, or reject the whole thing . . . I do think that the historical record of Christ's life on earth and his Resurrection is a very powerful one.*[7]

The success of the Human Genome Project was guaranteed at a Washington press conference in June 2000. From the podium, President Clinton told the world that human genome was "the language with which God created man." On one side of the president stood Francis Collins, leader of the NIH—or public—arm of the Human Genome Project. Collins listened to the president: by the summer of 2006, Collins's book *The Language of God*[8] was right up there with Ann Coulter on the bestseller list. On Clinton's other side in 2000 stood Craig Venter, head of Celera, the private company that set the pace of the quest. Venter had also paid attention to Clinton's phrase: he protested that "the human genome is not the book of life, it is not the blueprint of humanity, it is not the language of God. . . ."[9] The father of the project, Jim Watson, seated behind Clinton, Venter, and Collins on the podium, had no immediate comment on the language, syntax, or intentions of the deity. By the summer of 2006, Watson's magisterial book *DNA: The Secret of Life*[10] had not appeared on any bestseller list. At the time, Sidney Brenner quipped that "President Clinton described the human genome as the language with which God created man. Perhaps now we can view the Bible as the language with which man created God."[11]

Whether one holds with Collins or Venter, Watson or Brenner, we might suggest that modern experimental science is by definition, uni- and not multicultural. Scientists from many cultures and all sorts of religious belief work in a republic of reason, where each can claim *"Civis Scientium sum."* The threat to that republic is not the personal belief of any one scientist, but the intrusion of organized belief into the practice or teaching of science itself. Sad to say, faith-based fear of science, as documented by Miller, et al.,[12] in the case of evolution, has been around for a while.

DARWIN'S "AMERICAN COUSIN": JOHN WILLIAM DRAPER

John William Draper, M.D.
(1811–1882)

Faith-based fear of science has a long history in America. As late as 1854, doctors in New York State were not permitted to dissect the human body in medical schools, a practice routine in Europe for centuries. Finally, a statute called the "Bone Bill" passed the legislature by a whisker, but only after a bitter public battle between proponents of medical science and defenders of "natural law." The doctors pleaded that Vesalius's great dissection atlas, *De Humanis Corporis Fabrica*, had been used as a manual by medical students in more enlightened parts of the world since 1543. But in 1854, *Harper's Weekly* argued the anti-dissection cause in phrases that sound very much like the Bush administration's stand against stem-cell research or Governor Huckabee on "intelligent design":

> *Science may prove ever so clearly that there is nothing there but carbon, oxygen, and lime . . . but all this can never eradicate the sentiment we are considering, and that is too deeply in our laws of thinking, our laws of speech, our most interior moral and religious emotions.*[13]

Interior moral and religious emotions yielded to medical science when antebellum Manhattan finally caught up with Renaissance Brussels. The doctors' public campaign had carried the day: "Teach Anatomy, Learn Science," so to speak. The driving spirit behind the politics of the Bone Bill was John William Draper, founder and first president of New York University School of Medicine.[14] Draper sounded the tocsin for the progressive side in his 1874 *The History of the Conflict Between Religion and Science*, a bestseller that went through twenty editions in English alone:

The history of science is not a mere record of isolated discoveries. It is a narrative of the conflict of two contending powers—the expansive force of the human intellect on one side and the compression arising from traditional faith and human interest on the other. As large a number of persons now live to seventy years as lived to forty, three hundred years ago.[15]

We might note that mean longevity in the United States in 1840 was about forty years. By the 1980s, it was approaching eighty years.[16] That doubling of human longevity, as Draper would have been the first to argue, can hardly be credited to major increments in "our most interior moral and religious emotions."

Draper not only wrote a book that became the *machine de guerre* of free thought for the better part of a century, but as Professor of Chemistry at NYU, shared with Samuel F. B. Morse, Professor of Fine Arts at NYU, the honor of producing the first daguerreotype portraits by an American. In 1840, the two entrepreneurial professors opened up a commercial photography studio at the top of the University building in Washington Square; they also taught photography to the likes of young Mathew Brady. They collaborated successfully on another project at NYU in the 1840s: tinkering with an early version of his telegraph, Morse called on Draper to help design cables that might transmit electromagnetic signals over distances longer than that of their studio perimeter. Draper framed and tested the hypothesis that the conducting power of an electric wire varied directly with its diameter, an equation that made commercial telegraphy possible.[17] When in 1844 Samuel F. B. Morse transmitted the message "What hath God wrought?" from Washington to Baltimore, he might just as readily have asked, "What hath Draper wrought?"

In 1847 Draper published his "Production of Light by Heat," an early contribution to spectrum analysis. Indeed, he was the first to photograph the diffraction spectrum.[18] He was also the first to take a photograph of the moon and, with his son, to prepare the first photomicrographs of tissues. Ever the polymath, he became the featured lecturer at the historic British Academy debate at Oxford, the "Monkey Debate" of 1860, between the evolutionists, led by Thomas Huxley, and the religious, led by Bishop Wilberforce. Draper (a.k.a. "Darwin's American Cousin") had been invited as the American champion of Charles Darwin while the doyen of American biology, Louis Agassiz, was still flying the banner of creationism at Harvard.[19] Draper's topic was "The Intellectual Development of Europe Considered with Reference to the Views of Mr. Darwin." It marked the first entry of an American on the stage of evolution.

Looking back at that Victorian battlefield, Owen Chadwick, a Cambridge historian of religion, dismissed Draper as a shallow village atheist:

Draper's books contain the paean of praise to science, a hymn, its mighty achievements, among them the telegraph, telescopes, balloon, diving bells, thermometer, barometer, medicines, railway, air pump batteries, magnets, photographs, maps, rifles, and warships. . . . Draper never stopped to ask himself why anyone who invented a camera or possessed a barometer might be led to think his faith in the God of Christianity shaky.[20]

I'm convinced that while our twentieth-century minds may not be secularized by telescopes, balloons, diving bells, and thermometers—or scanning GenBank, for that matter—a glance at the bills of mortality might let us pay respect for Draper's kind of reductionist science. Indeed, I conclude that we're not living longer because of Bishop Wilberforce's belief system, but because of what medical science has learned over the years thanks to contributions like Draper's Bone Bill, his microscopy, and his Darwinism.

We're likely to live even longer in years to come. For that we can thank not only reductionists like Draper and Watson but also true believers such as Francis Collins. The republic of science has room for all. In fact, with its philanthropic thrust and social concerns, Collins's *The Language of God* could have used this nineteenth-century paean to religion as its epigraph:

Man is the world of man [and] religion is the general theory of that world, its encyclopedic compendium, its logic in popular form, its spiritual point d'honneur, its moral sanction, its solemn completion, its universal ground for consolation. . . . Religion is the sigh of the oppressed creature, the heart of a heartless world, the soul of a soulless condition.[21]

This passage from Karl Marx ends with the well-known phrase "It is the opium of the people." When we wake up from pipe dreams, we can teach evolution and learn science, together.

21. Diderot and the Yeti Crab: The Encyclopedias of Life

Denis Diderot (1713-1784), of the *Encyclopédie* (1751)

"Yeti Crab," *Kiwa hirsuta*, of the *Encyclopedia of Life* (2007)

> *Tantum series juncturaque pollet,*
> *Tantum de medio sumptis accedit honoris.*
>
> (So great is the power of linkage and order,
> That even the mundane becomes important.)
> —Epigraph (from HORACE) of Diderot's *Encyclopédie*[1]

Imagine an electronic page for each species of organism on Earth, available everywhere by single access on command. . . . Each species is a small universe in itself, from its genetic code to its anatomy, behavior, life cycle, and environmental role, a self-perpetuating system created during an almost unimaginably complicated evolutionary history. Each species merits careers of scientific study and celebration by historians and poets. —E. O. WILSON, *The Encyclopedia of Life*[2]

> *Why I think that there must be someone*
> *on top of that small speck of dust...*
> *He's alone in the universe!*
> *I'll just have to save him*
> *Because after all,*
> *A person's a person, no matter how small.*
> —DR. SEUSS, *Horton Hears a Who*[3]

PAIDEIA, AS IN ENCYCLOPEDIA

OUR NEW CENTURY HASN'T EXACTLY FULFILLED THE ENLIGHTENMENT dreams of reason and order. Our government can't build levees or insure the sick, even as it remains in thrall to Bible-thumpers who attack evolution as "atheistic theology posing as science."[4] On the other hand, there are signs that reason and order remain alive and well in modern biology.

Early in 2007—a decade and a half into the genomic era—two important genomes were decoded: that of a macaque monkey and that of James Watson. The macaque genome reflects twenty-five million years of evolution since the macaque's ancestors split from those of chimps and humans. Macaque DNA, sequenced at a cost of $20 million and published in *Science*,[5] differs no more than 7 percent from that of humans. The DNA of chimps, which split from our line six million years ago, differs by only 1 to 2 percent from that of humans. In Houston, James Watson was presented with a map of his own, unique genome that had been worked out by an academic/industrial group at a cost of $1 million. The results remain unpublished. Experts agreed that the cost to each of us for having our own genome mapped could soon drop to $1,000 per pop.[6]

It's clear that we face a formidable task if we want to link this mass of genomic information to human life, death, and disease. More formidable still would be to set this material in the context of the world we inhabit. We'd probably have to invent an instrument that could encompass all of this knowledge in one place. We'd probably call it an encyclopedia, as Denis Diderot did when he defined the publication that made the eighteenth century the Age of Reason:

ENCYCLOPEDIA, *noun, feminine gender (Philosophy). This word signifies unity of knowledge; it is made up of the Greek prefix* EN, *in, and the nouns* KYCLOS, *circle, and* PAIDEA, *instruction, science, knowledge.*[7]

A CREATURE'S A CREATURE, NO MATTER HOW SMALL

An equally ambitious voyage to *Paidea* was announced in 2007 when six major scientific institutions launched the Encyclopedia of Life (EOL).[8] The mission, discussed for years among biologists devoted to ecology and evolution,[9] is to create a web-based compendium, with one page for every living species on the planet. Eventually, we will all appear in the Encyclopedia: animals, plants, fungi, and microbes—each of us—on one website, at one click, and with open access to all. The Encyclopedia will not only assemble everything known about the 1.8 million species already named and cataloged, but also to help in the hunt for the perhaps 100 million species still out there waiting for binomial recognition.

While there are six founding institutions (the Smithsonian, the Field

Museum of Chicago, Harvard University, the Marine Biological Laboratory of Woods Hole, the Biodiversity Heritage Library, and the Missouri Botanical Garden), it's clear that the guiding spirit behind the work is Harvard's E. O. Wilson, who laid down the plan in 2003.[8] As work on the Encyclopedia progresses, Wilson expects that it will encounter the new and unexpected:

> *New classes of phenomena will come to light at an accelerating rate. Their importance cannot be imagined from our present meager knowledge of the biosphere and the species composing it. Who can guess what the mycoplasmas, collembolans, tardigrades, and other diverse and still largely unknown groups will teach us?*[11]

The pages already available on the Encyclopedia's website (www.eol.org) show us creatures great and small, known and lesser known, simple and complex. We find polar bears (*Ursus maritimus*) and rice sprouts (*Oryza sativa*), poison mushrooms (*Amanita phalloides*) and Yeti crabs (*Kiwa hirsuta*). The crabs have clusters of miniscule bacteria (unidentified so far) clinging to their claws. So new are the crabs to our ken, that when plans for the Encyclopedia were being formulated in 2005, the Kiwa family of crustaceans was unknown. Who, indeed, as Wilson asks, can guess what we'll learn from the bugs on the claws of the crab?

Each page of the Encyclopedia of Life will feature its own, carefully vetted table of contents that will link to a creature's evolutionary history, taxonomic description, pictures, maps, videos, sound, and sightings, as well as to its physiology, molecular biology, behavior, ecology, diseases, life span, etc. The first version of EOL will take about ten years to complete and is expected to fill about 300 million pages, which, if lined up end to end, would be more than 83,000 kilometers long, able to stretch twice around the world at the equator. The MacArthur and Sloan foundations have given $12.5 million to pay for the first two and a half years of the effort.[12] Like Wikipedia, the project will have open access. In sections reserved for the general public, amateur birders, naturalists, schoolchildren, and others will be able to contribute their bits; but unlike Wikipedia, all material to appear in the main, or "expert," section will be refereed by scientists before publication.[13]

The Scientific Advisory Board (with Wilson and Gary Borisy of the MBL as co-chairs) will supervise scientific sites around the world where the primary work will be done: scanning millions of pages, ordering data, setting appropriate links, and posting the material for publication on the web. Much of the information technology is already up and running, thanks in no small part to my colleagues at the MBLWHOI Library's uBio project (Cathy Norton, David Remsen, David Patterson, and Patrick Leary), who worked out methods for reconciling the global Babel of conflicting taxonomies and now have 10 million names (not species) under their belt.[14]

One species, one title page, may seem a fair allotment to the biologist; it remains to be seen whether critics like Senator Brownback will appreciate having humans (*Homo sapiens* Linnaeus, 1758) granted encyclopedic equality with the oriental cockroach, also first tagged by Linnaeus (*Blatta orientalis* Linnaeus, 1758). Senator Brownback insisted in the *New York Times* that "Man was not an accident and reflects an image and likeness unique in the created order."[4] And not *Blatta orientalis* or the Yeti crab?

VERITÉ AND THE EARTHWORM

The Encyclopedia of Life has a noble goal, however bland the prose of its mission statement:

> *To transform the science of biology, and inspire a new generation of scientists, by aggregating all known data about every living species. And ultimately, to increase our collective understanding of life on Earth, and safeguard the richest possible spectrum of biodiversity.*[16]

The EOL mission statement suffers somewhat by comparison with Denis Diderot's plan for his *Encyclopédie*:

> *In fact, the purpose of an encyclopedia is to gather knowledge from the four corners of the earth and to present it in an organized fashion to our contemporaries as well to those who will live after us; in order that the work of ages past will not become meaningless in the future, and that our successors, better educated thanks to this effort, will become happier and more ethical, and that we will not have died without having been of service to human kind.*[17]

Of course it all sounds better in French (. . . *que nous ne mourions pas sans avoir bien mérité du genre humain*)—and of course Diderot didn't have to submit his draft to a committee.

The *Encyclopédie* had its start in 1747, when Diderot and the Abbé Gua de Malves signed on with a commercial publisher for a new encyclopedia based on *Chambers' Encyclopedia* of London. Diderot had cut his teeth on the non-profitable *Dictionnaire de médecine* (1746). He now assembled his fellow *philosophes* including d'Alembert, Voltaire, Rousseau, and Montesquieu, as *La Société des Gens de Lettres*, to prepare and distribute the more ambitious *Encyclopédie* to subscribers, among them the gentry of Enlightenment Europe.[18] Eschewing royal societies and academies—as they slighted him— Diderot insisted that the work be carried out by "a society of men of letters and artists in order to assemble every talent. I will have them dispersed, because there is no existing society [i.e., royal societies] from which one can draw all the knowledge that is needed, and because, if you wanted the project to be forever in the making and never completed, you could do no better than to create such a society."

The first volume of the *Encyclopédie* appeared in 1751, and despite random interruptions by crown and censor, eventually reached thirty-five volumes. Diderot himself edited twenty-eight of these between 1751 and 1766. Diderot viewed his *Encyclopédie* as a "*machine de guerre*" aimed at the stiff hierarchies of his day, clerical and secular alike. He hoped that "both the man of the people and the scientist will always have equally as much to desire and instruction to find in an encyclopedia."[19]

He would have been delighted to hear our own encyclopedists proclaim, "The EOL is intended to be a bridge between science and society and between scientists and citizens, as well as a research environment for scientists."[20]

Frontispiece of the 1772 edition of the *Encyclopédie*.
Drawn by Charles-Nicolas Cochin, engraved by Bonaventure-Louis Prévost

A 1772 edition of the *Encyclopédie* sported a frontispiece that might well serve a similar function on the website of the Encyclopedia of Life. The engraving shows Truth (*Verité*) at the apex of the composition with Reason and Philosophy lifting her veil to reveal Enlightenment. *Verité* sheds her light equally on the various arts and crafts (on the left) and on the sciences, including mathematics, optics, and geometry (on the right). At the foot of *Verité* sits Theology, her back to the light, her eyes gazing at the clouds. The entire plan of the *Encyclopédie* was so structured as to cast theology in a supporting role to the

Verité of the useful Arts and Sciences.[21] The space allotted to "Divine Science" was not larger than that devoted to "The Manufacture and Uses of Iron."

Robert Darnton has pointed out that the ultimate triumph of the *philosophes* came when scholarly disciplines displaced parochial dogma during the nineteenth century. "But the key engagement took place in the 1750s, when the Encyclopedists recognized that knowledge was power and, by mapping the world of knowledge, set out to conquer it."[22] They did more; they sought diversity in nature. Diderot, author of a textbook of physiology and a devotee of natural science in general, went on to take a pre-Darwinian poke at creationism in his *Rêve d'Alembert*:

> . . . *you assume that animals were originally what they are now. What foolishness. We don't know any more about what they were than we do about what they'll become. The imperceptible earthworm which moves around in the mud is perhaps developing into the condition of a large animal, and an enormous animal, which astonishes us with its size, is perhaps developing into the condition of the earth worm and is perhaps a unique and momentary production of this planet.*[23]

Diderot got it right: the dinosaur line went bust, but the imperceptible earthworm went on to stand tall. A creature's a creature, no matter how small.

THE HAIRY GODDESS

Diderot's *Encyclopédie* was based on the notion that empirical descriptions of the here and now are preferable to the imaginings of metaphysics:

> *I have believed that the wing of a butterfly, well described, would bring me closer to divinity than a volume of metaphysics."*[24]

Diderot would have been reassured to find that the Encyclopedia of Life, another product of modern scholarship, pays careful attention to the here and now:

> *The Yeti crab is distinct from other related crab families in overall carapace morphology, leg morphology, vestigial eye and extraordinary setose nature of the claws. Examination of the setae (hairs) revealed several different types of bacteria which likely included sulphur-oxidizing strains.*

The discovery of the Yeti crab is not only a story with a French (and Woods Hole) connection, but reassures all of us that we can learn something new if we describe it well, be it butterfly wing or crab hair.

The crab was a serendipidous finding in the course of a 2005 expedition designed to discover how creatures found in deep hydrothermal vents in one part of an ocean can colonize other vents in vastly distant parts. To this end, an international team of marine biologists probed the ocean floor, 7,200 feet down, at a site 1,000 miles south of Easter Island in the Pacific. They were at the end of a six-hour dive in a deep submersible vehicle called *Alvin*, a craft best known for exploring the sunken *Titanic*,[25] when they struck scientific gold. Michel Segonzac, from the *Institut Français de Recherche pour l'Exploitation de la Mer* (IFREMER) in Brest, noticed unusually large, half-foot-long albino creatures in areas where warm water from geothermal vents was seeping into the ocean floor. The pilot of the *Alvin*, Anthony Tarantino of the Woods Hole Oceanographic Institution (where *Alvin* was developed) suctioned one of the blind, lobsterlike creatures into the vehicle by means of a vacuumlike hose known as the "slurp gun."[26]

It took less than a year of morphologic and molecular genetic analysis for Segonzac and his associates to ascertain that they had found not only a new species of an unknown genus, but—bigger still in the world of taxonomy —an entirely new family of crustaceans. They called the new family *Kiwaidae* (from Kiwa, the goddess of shellfish in Easter Island mythology), and its new Latin name became *Kiwa hirsuta*. But among marine biologists, the crab immediately became known as the "Yeti" crab, after the hirsute snowman of Himalaya legend. It seemed better than calling the creature a hairy goddess.

The data were rushed into print in the major French journal of taxonomy, *Zoosystema*,[27] and soon fell into taxonomic order on uBio.[28]

Kiwa hirsuta
Animalia
  Arthropoda
    Crustacea Brünnich, 1772
     Malacostraca Latreille, 1802
      Eumalacostraca Grobben, 1892
       Eucarida Calman, 1904
        Decapoda Latreille, 1802
         Pleocyemata Burkenroad, 1963
        Anomura Macleay, 1838
         Galatheoidea Samouelle, 1819
          Kiwaidae Macpherson, Jones and Segonzac, 2005
          Kiwa Macpherson, Jones and Segonzac, 2005
           Kiwa hirsuta Macpherson, Jones and Segonzac, 2005

The biology and evolutionary history of the Yeti crab will be forever linked in the Encyclopedia of Life to other living things with whom it shares our fragile planet, including *Homo sapiens*. Linkage and order, as in the *Encyclopédie*'s epigraph, make the hairy goddess one of those creatures, "created during an almost unimaginably complicated evolutionary history. Each species merits careers of scientific study and celebration by historians and poets."

Merci, Messrs. Wilson et Diderot.

22. Dengue Fever in Rio: Macumba versus Voltaire

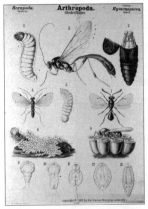

Arthropoda, by Rudolph Leuckart   Voltaire (1689–1778)

BRAZIL: DENGUE TOLL HITS 92 IN RIO STATE

The death toll from a dengue outbreak in Rio de Janeiro state has reached 92, topping what was previously the state's deadliest bout with the disease, Brazil's government news agency said Wednesday. Another 96 possible dengue fatalities have been reported in the current outbreak but have yet to be confirmed.
—*Washington Post,* April 24, 2008[1]

Meanwhile, Rio Mayor César Maia, recently on the road in northeast Brazil, prayed to a local [Macumba] god to carry the dengue mosquito out to sea.
—*Newsweek,* April 14, 2008[2]

"Yes, sir," [replied the Negro to Candide] "When we work in the sugar mill and get our fingers caught in the grinding machine, they chop off our hand; when we try to run away, they cut off a leg. That cost me an arm and a leg: it's the price of your eating sugar in Europe! And yet when my mother sold me for ten silver pieces on the coast of Guinea, she said to me, 'My dear child, bless our local Fetish gods; worship them forever; they will make you happy; you've been given the honor to be a slave to our lords, the white folk."
—VOLTAIRE, *Candide*[3]

DENGUE IN BRAZIL

I N THE MOST SEVERE EPIDEMIC EVER TO SWEEP THE COUNTRY, NEARLY 230,829 Brasilieros came down with dengue fever between January and April of 2008. As the death toll from dengue and dengue hemorrhagic fever neared 100, Brazil's Minister of Health, Jose Gomes Temporão, conceded that the country has "lost its war" against the disease, which had also broken out in 1986, 1995, and 2002. He warned that Brazilians would have to "coexist" with the disease for the foreseeable future.[4]

Thanks to effective insecticide measures, dengue had been largely eradicated from much of Latin America between 1950 and the early 1980s. But, as those efforts faded, and with DDT declared off limits, the mosquitoes reappeared. Last year, over a million people in the Americas became infected by dengue, more than half of them in Brazil, and 70 percent of those in Rio de Janeiro. In Rio itself, the neighborhoods hit the hardest every year have been its hillside slums, the *favelas*.[5]

In 2008, *Aedes aegypti*, the mosquitoes that transmit dengue in Brazil, became a vast insect horde. They propagated en masse in the stagnant gutters, drains, and puddles left by the record-setting rains of the Brazilian summer of 2007. Nurture was added to nature. Superstitious slum dwellers and the drug lords who run *favelas* such as the teeming *Cidade de Deus* (City of God) prevented uniformed sanitary workers from entering their fiefdoms. Sadly, children under fifteen accounted for almost half of the deaths from dengue in Rio.[6]

THE VIRUS AS PATHOGEN

The epidemiology and pathophysiology of the dengue virus have been well studied. Dengue virus is a genus of the Flaviviridae family, which consists of more than sixty-eight members and includes the West Nile virus and that of yellow fever. Most flaviviruses are borne by persistently infected arthropod vectors that go on to infect their vertebrate hosts. The enveloped viral particle of a flavivirus contains a single-stranded, positive-sense RNA genome of approximately 11 kb, whose replication is primarily cytoplasmic and membrane associated.[7] A variety of mosquitoes carry the four serotypes of dengue virus (D1–D4), which differ only slightly in their capacity to cause dengue fever, dengue hemorrhagic fever (DHF), or dengue shock syndrome. A first infection by dengue virus is usually mild, almost flulike; but the severe forms such as "breakbone fever," with its triad of fever, rash, and acute muscle pain, usually follow the second exposure to the same or any other serotype. The crueler syndromes that lead to death—dengue hemorrhagic fever and blood-loss shock—tend to affect younger patients, with mortality rates up to 60 percent.[8]

The disaster of dengue hemorrhagic fever is caused in large part by immune complexes in which viral antigens are tagged by our own antibodies.[9] These in turn provoke a coagulum in the bloodstream, consisting of platelets and white cells; when such aggregates get stuck in the smaller blood vessels of lung or kidney, organ damage ensues. The blood vessels are themselves injured by a witches' brew of chemical messengers of inflammation.

Acute respiratory failure follows as both bleeding and clotting are deranged. Blood vessels leak fluid, the blood that is left in vessels becomes concentrated, and—alas—the patient finally bleeds to death as the vessels lose their fight with the virus.[10] There is no specific treatment other than hydration and/or transfusion; steroids are useless.

MACUMBA AND THE FLYING SYRINGE

Since dengue fever and DHF cannot be adequately treated, only preventive measures can quell an epidemic. Alas, in Rio that year, politics stood in the way of prevention. The doctors' union accused the local government of criminal negligence in protecting public health. The doctors complained that Cesar Maia, the mayor of Rio, refused to appeal to the federal government for sufficient public-health workers. Instead, the mayor trekked to a mountaintop in Bahia to plead with a local Macumba deity to rid the country of mosquitoes. Issuing a statement that may sound familiar to North American observers, the president of the doctors' union protested that "the threat of an epidemic was already apparent since last year and the city did nothing. The mayor can't run for re-election, so he left it all to luck and the Lord."[11] Or, as they put it in Portuguese, *"Enquanto o digníssimo Prefeito da Cidade do Rio vai fazer macumba na Bahia pra espantar o Aedes Aegypti."* (The estimable mayor practiced Macumba in Bahia to exorcise the mosquito.)[12]

Macumba—the name is of West African origin—describes a religion that mixes African fetish worship, Brazilian spiritualism, and Roman Catholicism in various proportions. The original African fetish gods, or Orixás, became equated with their corresponding Christian saints and are now equally honored. From humble beginnings in the slave ships of the eighteenth century, the sacraments of the two main sects of Macumba, the Candomblé and Umbanda, have become quite theatrical. The ceremonies feature ritual blood sacrifice (mainly of crowing cocks), scented candles, burning incense, and buckets of fresh flowers. They are led by mediums who communicate with holy spirits by falling prostate before outdoor altars surmounted by the sign of the Cross.[13] As dengue decimated the *favelas* of Brazil this year, cries of the cock and aromas of wax filled the air. What was missing was a whiff of insecticide.[14]

(Since the slave trade cut a wide swath across the Americas, the Caribbean basin and Central America became centers of a similar brand of synthetic religion called Santeria. After the Cuban exodus of the early 1960s, Santeria entered the U.S. on the coast of Florida, where its practice has run into difficulties over those crowing cocks.[15])

Central America and the Caribbean basin have also become a vast reservoir for the dengue virus; outbreak after virulent outbreak has prompted major U.S. aid.[16] Over the last twenty years, serotype circulation in the region has gone from none or single to multiple, and the mosquitoes are moving north. Our porous southern border has permitted dengue to become endemic in southern Texas, and the disease is now moving into California.[17] "Mosquitoes are flying syringes," warned Dr. Cortez-Flores of Loma Linda School of Public Health, an expert on the arthropods.[18] Syringes know no borders.

The notion, I might add, will be familiar to students of the slave trade and its cruelties. *Aedes aegypti,* a prime vector not only of dengue but also of yellow fever, reached South America in slave ships on which the mosquitoes' eggs survived in water containers while live mosquitoes fed on the helpless Africans and sailors on board. Michael Nathan, a WHO entomologist, noted that "Aa lot of slaves and a lot of crews died of yellow fever on the way over."[19]

MACUMBA AND THOROUGHWORT

Sandro Cesar, president of the Rio de Janeiro state health workers' union, accused Rio's mayor (Cesar Maia, that fan of Macumba) of turning down an offer of some 3,200 federal sanitary aides, because of his political dispute with the federal government. The mayor defended himself to the AP: "That's a lie," he said and blamed political agitators "who don't want to work" for spreading false accusations against him.[20]

But the fight against dengue gained support from other quarters. The British supermodel Naomi Campbell stretched a white "Rio Against Dengue" T-shirt across her poitrine to lead a drive for blood donation.[21] Others took their cue from Dr. Dráuzio Varela, a TV practitioner, who had gained national attention by suggesting that the simple addition of a spoonful of soup to stagnant waters would instantly kill any insect larvae present: "Better than DDT!" he told viewers.[22] Offering dilutions of grandeur, a team of healers from the Homeopathic Action Group for Humanitarian Aid (*Homeopatia Ação Pelo Semelhante,* an NGO) offered their services to forty children at a state hospital in Rio. Erroneously, the well-meaning homeopaths asserted that "the faster the disease comes on, the faster is the recovery."[23] One finds that the treatment recommended by homeopaths for dengue fever includes *Eupatorium perforatum* (thoroughwort) and white briony at almost infinite dilutions.[24]

These unorthodox measures proved futile. Owing to her medical history, the supermodel's offer of blood was refused and an entomologist of the eminent Oswaldo Cruz Institute went before TV to show that Dr. Varela's spoonful of soup had absolutely no effect on the viability of *Aedes aegypti* larvae.[25] And as for *Homeopatia Ação,* a controlled, randomized study had already shown that 10^{24} dilutions of thoroughwort and other herbs were no more effective than lactose in the treatment of dengue fever. Large names, little herbs, with little efficacy: Aconita, Bryonia, *Eupatorium perforatum*, Gelsemium, and *Rhus toxicodendron*.[26] Those charming names could have decorated the lyrics of Leonard Bernstein's *Candide*, in which a doctor offers remedies for the ills of the Lisbon earthquake:

> *Here be powders and pills*
> *For your fevers and chills.*
> *I've a cure safe and sure*
> *For whatever your ills.*
> *For a fit of migraine,*
> *Or a pox on the brain*
> *Here's an herb that will curb any pain!*[27]

CANDIDE IN ELDORADO

Voltaire was prompted to write *Candide* (1759) in response to the great Lisbon earthquake of November 1, 1755, a disaster that killed thousands of innocents, including many children. The disaster literally shook the philosopher's faith in divine providence forever:

> *After the earthquake, which had destroyed three-fourths of the city of Lisbon, the sages of that country could think of no means more effectual to preserve the kingdom from utter ruin than to entertain the people with an auto-da-fé, it having been decided . . . that the burning of a few people alive by a slow fire, and with great ceremony, is an infallible preventive of earthquakes.*[28]

"*Auto-da-fé*," I might add, is Portuguese for "act of faith," and to my mind would cover both the burning of heretics by the Inquisitors in Lisbon and the mayor's appeal to Orixás after the dengue epidemic in Rio. "What a day, What a day, for an *auto-da-fé*" they sing in Bernstein's musical; what a day, what a day to blame someone else. If syringes know no borders, natural disasters know no end of blame. The mayor blamed outside agitators, the doctors blamed the mayor, journalists blamed the drug lords while Brazilian nationalists blamed Mexicans and Caribbean islanders for permitting mosquitoes to flourish in the Americas.[29] Voltaire blamed credulity.

Candide, having witnessed earthquake, mutilations, and that Portuguese *auto-da-fé*, recalls the pablum fed him by his mentor, Dr. Pangloss, and asks the proper question:

> *Candide, amazed, terrified, confounded, astonished, all bloody, and trembling from head to foot, said to himself, "If this is the best of all possible worlds, then what are the others?"*[30]

Voltaire transports Candide to Latin America to try out life in one of those possible New Worlds and to find his lost love. He encounters unlettered natives, overworked slaves, and colonial rascals; he finds corruption and sloth are as rampant in South America as in Europe. And then, he's off to a land that all of us dream about: Eldorado.

ORDEM E PROGRESSO

Today, Brazil may not be Eldorado, but it does contain a society as broadly tolerant, diverse, and multiracial as any on earth. Macumbans worship next to Roman Catholics. Black and white, Indian and Iberian, Candomblé and Umbanda sun themselves and make love on the sands of Ipanema; they also spawn those amazing soccer players. The country fulfills Voltaire's prophecy that "if a land has only one religion, it tends to despotism; if it has two, they are at each others' throats. If it has thirty, they live together in peace."[30] When slavery in Brazil was abolished in the 1880s by a republic that overthrew its monarch,[31] the country adopted a motto on its flag that remains there today. "Ordem et Progresso." Order and Progress are no shabby principles to sport on any country's flag these days. The motto is adapted from August Comte, the French physician and positivist, whose *Plan de travaux scientifiques nécessaires pour réorganiser la société* (1822) was based on Voltaire's notion of society based on religious tolerance and reason, i.e., order and progress. Comte's plan for a just polity added fraternal love as a first principle: **L'AMOUR POUR PRINCIPE ET L'ORDRE POUR BASE; LE PROGRES POUR BUT** (With love as principle, order as foundation, and progress its aim).[32] Comte acknowledged that he was much in debt to Voltaire's wandering optimist, Candide, who found his Eldorado in a fictional Brazil where:

> *Candide asked to see the High Court of Justice, the Parliament; but was answered that they had none in that country, being utter strangers to lawsuits. He then inquired if they had any prisons; they replied none. But what gave him at once the greatest surprise and pleasure was the Palace of Sciences, where he saw a gallery two thousand feet long, filled with various devices of mathematics and natural science.*[33]

That gallery came to life a century ago when Brazil founded its Instituto Oswaldo Cruz, to put "science at the service of the Brazilian people." Its scientists busy themselves with the newest devices of molecular biology to map the dengue virus and make its diagnosis neat and rapid.[34] Finally, as this was written, the federal government had overridden the mayor and was moving its battalions of health workers into Rio to kill mosquitoes and cleanse the *favelas*.

Acknowledgments

T HESE ESSAYS ARE EXTENSIONS OF MATERIAL THAT HAS PREVIOUSLY appeared in the *FASEB Journal* (the official *Journal of the Federation of American Societies for Experimental Biology*), a publication for which I am responsible. I am in sincere debt to the remarkable staff at the FASEB Publications Office in Bethesda, and could not have carried out that responsibility without the expert advice and editorial savvy of Jennifer Pesanelli, Cody Mooneyhan, Susan Moore, and their associates. At Woods Hole, I have been quartered in the MBL/WHOI library, a living archive of science and learning directed by Cathy Norton. I am grateful for her help, for that of her staff, and for making the MBL a place where ideas spring not only from books and the web, but also from conversation. My major editorial debt, as always, is to the patient counsel, critical eye, and unflagging spirit of Andrea Cody, the Administrative Coordinator of the Biotechnology Study Center at the NYU School of Medicine; and without Erika Goldman of the Bellevue Literary Press, these essays would have remained *in silico*. *Sans cela, rien!*

References

PREFATORY NOTE

1. C. P. Snow, Two Cultures and the Scientific Revolution. (Cambridge: Cambridge University Press, 1959), p.11
2. W. H. Auden, "The Art of Poetry" (interview). *Paris Review* 8 (1947), pp.1–37.

MORTAL AND IMMORTAL DNA: CRAIG VENTER AND KEATS'S "LAMIA"

1. J. Craig Venter Institute, "First Individual Diploid Human Genome Published by Researchers at J. Craig Venter Institute/Sequence Reveals That Human to Human Variation Is Substantially Greater Than Earlier Estimates," http://www.jcvi.org/press/ (accessed October 2007).
2. G. Cossu, and S. Tajbakhsh, "Oriented Cell Divisions and Muscle Satellite Cell Heterogeneity," *Cell* 129, no. 5 (June 2007), pp. 859–861.
3. Lewis Thomas, *The Medusa and the Snail: More Notes of a Biology Watcher* (New York: Bantam, 1979), p. 28.
4. Nobel Foundation, "The Nobel Prize in Physiology or Medicine 2007: Press Release," http://nobelprize.org/nobel_prizes/medicine/laureates/2007/press.html (accessed October 2007).
5. W. Gilbert, "The RNA World," *Nature* 319 (February 20, 1986), p. 618.
6. Nobel Foundation, "The Nobel Prize in Chemistry 1989: Press Release," http://nobelprize.org/nobel_prizes/chemistry/laureates/1989/press.html (accessed October 2007).
7. J. Couzin, "Breakthrough of the Year: Small RNAs Make Big Splash," *Science* 298 (December 20, 2002), pp. 2296–2297.
8. Nobel Foundation, "The Nobel Prize in Physiology or Medicine 2007: Press Release," http://nobelprize.org/nobel_prizes/medicine/laureates2006/press.html (accessed October 2007).
9. B. Orelli, "A Pharma Divorce, and Good Riddance," *The Motley Fool* (September 20, 2007), http://www.fool.com/investing/small-cap/2007/09/20/a-pharma-divorce-and-good-riddance.aspx (accessed October 2007).
10. J. Watson, *DNA: The Secret of Life* (New York: Knopf, 2003).
11. G. Weissmann, *The Year of the Genome.* (New York: Times Books, 2002).
12. T. A. Rando, "The Immortal Strand Hypothesis: Segregation and Reconstruction," *Cell* 129 June 29, 2007), pp. 1239–1243.

13. S. Levy, et al., "The Diploid Genome Sequence of an Individual Human," *PLoS Biol.* 5, no. 10, (October 2007), e2113–2144.

14. R. Weiss, "Mom's Genes or Dad's? Map Can Tell: One Man's DNA Shows We're Less Alike Than We Thought," *Washington Post,* September 4, 2007, p. A01.

15. Prix Galien USA, http://www.prix-galien-usa.com/ (accessed October 2007).

16. J. Cairns, "Mutation Selection and the Natural History of Cancer, *Nature* 255, no. 5505 (May 15, 1975), pp. 197–200.

17. Rando, "The Immortal Strand Hypothesis."

18. P. M. Lansdorp, "Immortal Strands? Give Me a Break," *Cell* 129, pp. 1244–1247.

19. J. R. Merok, et al., "Cosegregation of Chromosomes Containing Immortal DNA Strands in Cells That Cycle with Asymmetric Stem Cell Kinetics," *Cancer Res.* 62, pp. 6791–6795.

20. P. Karpowicz, et al., "Support for the Immortal Strand Hypothesis: Neural Stem Cells Partition DNA Asymmetrically *in Vitro,*" *J. Cell Biolology,* 170, pp. 721-732.

21. V. Shinin, et al., "Asymmetric Division and Cosegregation of Template DNA Strands in Adult Muscle Satellite Cells. *Nature Cell Biol.* 8, pp. 677–687.

22. "DNA Replication: Avoid All Counterfeits," *CNRS International Magazine,* http://www2.cnrs.fr/en/694.html (accessed October 2007).

23. Nobel Foundation, "The Nobel Prize in Physiology or Medicine 2007."

24. Levy, et al., "Diploid Genome Sequence."

25. J. Keats, "Lamia," *Poetry and Letters of John Keats,* ed. Elizabeth Cook (New York: Oxford University Press, 1990), p. 305.

26. Ibid.

27. Ibid.

28. Levy, et al., "Diploid Genome Sequence."

HOMEOPATHY: HOLMES, HOGWARTS, AND THE PRINCE OF WALES

1. O. W. Holmes, *The Autocrat of the Breakfast Table* (Boston: Houghton Mifflin, 1892), p. 40.

2. S. Hahnemann, *The Organon of the Rational Art of Healing,* http://www .homeopathyhome.com/reference/organon/organon.html (accessed July 2006).

3. Ibid.

4. Ibid.

5. R. English and J. Hope, "Charles at War with Doctors; Don't Close Your Mind to Alternative Therapy Says Prince," *Daily Mail,* May 24, 2006.

6. A. Cowell, A. "Favorite Theme of Charles Earns Pre-emptive Riposte," *International Herald Tribune,* May 26, 2006.

7. B. J. Feder, "More Britons Trying Holistic Medicine," *New York Times,* January 9, 1985, p. C1.

8. C. Blackstock, "Prince Orders Cost Study of Alternative Medicine," *The Guardian,* August 24, 2005, p. 4.

9. M. Baum, et al., "Re Use of 'Alternative' Medicine in the NHS," Thetimes.com, http://www.timesonline.co.uk/article/0,,8122-2191985,00.html (accessed July 2006).

10. A. Shang, et al., "Are the Clinical Effects of Homeopathy Placebo Effects? Comparative Study of Placebo-Controlled Trials of Homoopathy and Allopathy," *Lancet* 366, pp. 726–732.

11. English and Hope, "Charles at War."

12. "Charles Promotes Alternative Healthcare for Alzheimer's Disease," *Liverpool Daily Post,* May 23, 2006, p. 2.

13. D. Hencke and R. Evans, "Charles to Finance U.S. Ageing Research: British Experts Query Grant to Alternative Medicine Centre," *The Guardian,* September 6, 2006, p. 9.

14. National Center for Complementary and Alternative Medicine, "Energy Medicine: An Overview," http://nccam.nih.gov/health/backgrounds/energymed.html (accessed July 2006).

15. National Center for Complementary and Alternative Medicine, "Research Report: Questions and Answers About Homeopathy." http://nccam.nih.gov/health /homeopathy/html (accessed July 2006).

16. A. S. Relman, "A Trip to Stonesville," *New Republic* 219, pp. 28–38.

17. N. M. Hadler, "Fibromyalgia and the Medicalization of Misery," *J. Rheumatol,* 30, pp. 1668–1670.

18. I. R. Bell, et al., "Improved Clinical Status in Fibromyalgia Patients Treated with Individualized Homeopathic Remedies versus Placebo," *Rheumatology* 43, pp. 577–582.

19. O. W. Holmes, Preface to *Homeopathy and its Kindred Delusions.* In: *Medical Essays,* vol. X of *The Standard Edition of The Works of Oliver Wendell Holmes* (Boston: Houghton Mifflin, 1883), p. 9.

20. "Principles of Homeopathy," http://holistic-online.com/Homeopathy/homeo _principles.html (accessed July 2006).

21. NCCAM, "Research Report."

22. M. Rouzé, "*Oscillococcinum,* Le Joli Grand Canard," *Science et Pseudo-sciences, Cahiers bimestriels de l'Association Française pour l'Information Scientifique* 202 (accessed July 2006).

23. S. Barrett, "Homeopathy: The Ultimate Fake," *Quackwatch.org,* http://www .quackwatch.org/01QuackeryRelatedTopics/homeo.html (accessed July 2006).

24. E. Lewis, "Dana Ullman: Treating Children with Homeopathic Medicines," *Hpathy Ezine,* http://www.hpathy.com/interviews/danaullman.asp (accessed July 2006).

25. "About Dana Ullman," Penguin Group USA, http://us.penguingroup.com /nf/Author/AuthorPage/0,,1000040566,00.html (accessed July 2006).

26. K. Garsombe, "Alternative Remedies for Anthrax," *Alternet.org.,* http://www .alternet.org/envirohealth/11814 (accessed July 2006).

27. "Integrative Therapies Program for Children with Cancer: About Us," Columbia University Medical Center, http://www.integrativetherapiesprogram.org /about /staff.php (accessed July 2006).

28. Harvard Medical School Osher Institute, http://www.osher.hms.harvard.edu/ (accessed July 2006).

29. Rosenthal Center for Complementary and Alternative Medicine, http://www .rosenthal.hs.columbia.edu/Anniversary_awards.html (accessed July 2006).

30. Holmes, *Medical Essays,* p. 319.

CITIZEN PINEL AND THE MADMAN AT BELLEVUE

1. E. Konigsberg and A. Farmer, "Father Tells of Slaying Suspect's Long Ordeal with Mental Illness," *New York Times,* February 20, 2008, p. B1.

2. P. A. Pinel, *Treatise on Insanity: In Which Are Contained the Principles of a New and More Practical Nosology of Maniacal Disorders than Has yet Been offered to the Public,* tr. D. D. Davis (Sheffield: W. Todd/Cadell and Davies, 1806), p. 21.

3. S. Freud, "Charcot." In: *Collected Papers,* tr. J. Barnays (London: Hogarth Press, 1956). p. 17.

4. Editorial, "Letting Madmen Roam." *New York Post,* February 21, 2008), p. 22.

5. Mike Eliot, "You Must Act," *New York Daily News,* February 21, 2008, p. 28.

6. Ibid.

7. "Is Murder the Threshold?" *New York Civic,* February 22, 2008, http://www.nycivic.org/.

8. Transcript, "Shrink Slay Suspect's Assault Rap," *New York Post,* February 19, 2008, p. 1.

9. Konigsberg and Farmer, "Father Tells."

10. A. Berenson, "Daring to Think Differently About Schizophrenia," *New York Times,* February 24, 2008.

11. J. Cummings, "City of Future Splits Over Homeless," *New York Times,* October 13, 1987, p. A26.

12, E. J. Tanay, "Homicidal Behavior in Schizophrenics," *Forensic Sci.* 32, no. 5 (September 1987) pp. 1382–1388.

13. H. Schanda, "Homicide and Major Mental Disorders: A 25-Year Study," *Acta Psychiat. Scand.,* 110, no. 2 (August 2004), pp. 98–107.

14. M. Eronen, P. Hakola, and J. Tiihonen, "Mental Disorders and Homicidal Behavior in Finland, *Arch. Gen. Psychia.,* 53, no. 6 (June 1996), pp. 497–501.

15. J. Tiihonen, et al., "Risk of Homicidal Behavior among Discharged Forensic Psychiatric Patients," *Forensic Scie. Interna.,* 79, no. 2 (May 31, 1996) pp. 123–129.

16. G. Weissmann, *The Woods Hole Cantata* (New York: Dodd, Mead), p. 26.

17. Pinel, *Treatise,* p. 89.

18. Pinel, *Treatise.*

19. P. Pinel, *Nosographie Philosophique: ou, la Méthode de L'analyse Appliquée a la Medicine* (Paris: J. A. Brosson, 1789).

20. E. R. Kandel, "Disorders of Thought and Volition: Schizophrenia." In: *Principles of Neural Science,* E. R. Kandel, J. H. Schwartz and T. M. Jessell (New York: McGraw-Hill, 2007), pp. 1188–1208.

21. Pinel, *Treatise,* p. 122.

22. S. J. Kile, "Neuropsychiatric Update: Neuroimaging Schizophrenia," *Psychopharmacol. Bull.,* 40, no. 4 (2007), pp. 156–167.

23. Freud, "Charcot."

24. R. W. Chambers, "The Bicetre in 1792." In: R. W. Chambers, *Edinburgh Journal* (London: Chambers and Orr, 1849), p. 169.

25. Pinel, *Treatise,* p. 89.

26. Kandel, "Disorders."

27. C. A. Tamminga and J. M. Davis, "The Neuropharmacology of Psychosis," *Schizophrenia Bull.* 33, no. 4, pp. 937–946.

28. M. Foucault, *Madness and Civilization*, tr. Richard Howard (New York: Pantheon, 1965), p. 57.

29. Ibid., p. 248.

30. Tamminga and Davis, "Neuropharmacology."

31. Pinel, *Treatise*, p. 171.

THE EXPERIMENTAL PATHOLOGY OF STRESS:
HANS SELYE TO PARIS HILTON

1. H. Selye, "Forty Years of Stress Research: Principal Remaining Problems and Misconceptions," *Can. Med.Assoc. J,* 115 (1976), pp. 53–56.

2, "Paris Hilton's Stress Starvation," *Boston Globe,* http://www.boston.com/ae/celebrity /articles/2007/05/31/paris_hiltons_stress_starvation/ (accessed July 2007).

3. D. J. Taylor, "The Ballad of the Los Angeles Century Regional Detention Centre; Verses on the Incarceration of Paris Hilton," *The Independent on Sunday,* June 10, 2007, p. 1.

4. D. Frosch, "Fighting the Terror of Battles That Rage in Soldiers' Heads," *New York Times,* May 13, 2007.

5. Paris Hilton, interviewed by Larry King, *Larry King Live,* CNN, June 28, 2007.

6. "She's a Celebuconvict No More: Hilton Is Free—To Eat Cake, Get Extensions," *Chicago Tribune,* June 27, 2007, p. 19.

7. Paris Hilton, *Larry King Live.*

8. C. W. Hoge, et al., "Combat Duty in Iraq and Afghanistan, Mental Health Problems, and Barriers to Care," *New Engl. J. Med.* 351 (2007), pp. 13–22.

9. "Mistreated Casualties; Veterans with Psychological Wounds Are Getting Shabby Treatment from the Department of Veterans Affairs, *Washington Post,* June 19, 2007, p. A16.

10. House Committee on Veteran's Affairs, "Statement of Stefanie E. Pelkey before the Committee on Veteran Affairs, House of Representatives," http://veterans.house .gov/hearings/schedule109/jul05/7-27-05f/spelkey.html (accessed July 2007).

11. E. Emery, "Ex-GI Diagnosed with PTSD Dies in 2-Car Collision on I-25," *Denver Post,* February 11, 2007, p. C2.

12. American Psychiatric Association, "DSM-IV-TR Criteria for PTSD," *Psychiat. News,* 37 (2007), p. 25.

13. R. Yehuda and A. C. McFarlane, "Conflict Between Current Knowledge about Posttraumatic Stress Disorder and Its Original Conceptual Basis," *Am. J. Psychiat.* 152 (1995), pp. 1705–1713.

14. H. Selye, *The Physiology and Pathology of Exposure to Stress* (Montreal: Acta, 1959).

15. A. Patmore, *The Truth About Stress* (London: Atlantic Books, 2006).

16. American Institute of Stress, "Stress, Definition of Stress, Stressor, What Is Stress? Eustress?" http://www.stress.org/topic-definition-stress.html (accessed July 2007).

17. Ibid.

18. http://www.frankloesser.com/work/theatre/2 (accessed July 2007).

19. "Manhattan Movie Sound Bites," http://www.rosswalker.co.uk/movie_sounds /manhattan.html *Frank Loesser Enterprises website* (accessed July 2007).

20. Patmore, *The Truth About Stress.*

21. H. Heckle, "Spanish Hotel Lets Guests Smash Rooms," Associated Press, July 3, 2007.

22. World Travel Guide, "The World War II Walking Tour of Paris," http://www.affiliate

.viator.com/brochure/product_show.jsp?CODE=3588WWII&ID=1010&PRODUCTID =1016&AUID=2672 (accessed July 2007).

23. S. Cavin, "World War II Never Ended in My House. Interviews of 12 Office of Strategic Services Veterans of Wartime Espionage on the 50th Anniversary of WW II," *Ann. N.Y. Acad. Sci.* 1071 (2006). pp. 463–471.

24. G. Wheler, et al. "Cortisol Production Rate in Posttraumatic Stress Disorder," *J. Clin. Endocrinol. Metab.* 91 (2006), pp. 3486–3489.

25. H. Selye, "A Syndrome Produced by Diverse Nocuous Agents," *Nature* 138 (1936), p. 3.

26. H. Selye, *The Stress of My Life* (New York: Van Nostrand Reinhold, 1979).

27. H. Selye, "Anesthetic Effect of Steroid Hormones," *Proc. Soc. Exptl. Biol. Med.* 46 (1941), pp. 116–121.

28. H. Selye, "Pharmacological Classification of Steroid Hormones," *Nature* 148 (1941), pp. 84–85.

29. H. Selye, G. Gentile, and P. Prioreschi "Calciphylaxis: Cutaneous Molt Induced by Calciphylaxis in the Rat," *Science* 134 (1961), pp. 1876–1877.

30. Ibid.

31. H. Selye, "The General Adaptation Syndrome and the Diseases of Adaptation," *J. Clin. Endocrinol. Metab.* 6 (1946), pp. 117–230.

32. W. B. Cannon, "Stresses and Strains of Homeostasis," *Am. J. Med. Sci.* 189 (1935), pp. 1–4.

33. W. B. Cannon and J. R. Pereira, "Increase of Adrenal Secretion in Fever," *Proc. Natl. Acad. Sci. USA.* 10 (1924), pp. 247–248.

34. Cannon, "Stresses and Strains."

35. H. Selye and C. Fortier "Adaptive Reaction to Stress," *Psychosomat. Med.* 12 (1950), pp. 149–157.

36. *Oxford English Dictionary* (New York: Oxford University Press).

37. D. Bird, "Dr. Hans Selye Dies in Montreal: Studied Effects of Stress on Body," *New York Times*, October 22, 1982, p. B10.

38. Selye, *The Stress of My Life.*

39. Ibid.

GORE'S FEVER AND DANTE'S *INFERNO*: CHIKUNGUNYA REACHES RAVENA

1. A. Gore, "Nobel Lecture, December 10, 2007," http://nobelprize.org/nobel _prizes/peace/laureates/2007/gore-lecture_en.html (accessed January 8, 2008).

2. P. R. Epstein, "Chikungunya Fever Resurgence and Global Warming," *Am. J. Trop. Med. Hyg.* 76 (2007), pp. 403–404.

3. E. Rosenthal, "As Earth Warms Up, Tropical Virus Moves to Italy," *New York Times,* December 23, 2007, p. A16.

4. G. Rezza, et al., "Infection with Chikungunya Virus in Italy: An Outbreak in a Temperate Region," *Lancet* 370 (2007), pp. 1840–1846.

5. P. J. Mason and A. J. Haddow, "An Epidemic of Virus Disease in Southern Province, Tanganyika Territory, in 1952–53: An Additional Note on Chikungunya Virus Isolations and Serum Antibodies," *Trans. Roy. Soc. Trop. Med. Hyg.* 51 (1957). pp. 238–240.

6. T. E. Morrison, et al., "Complement Contributes to Inflammatory Tissue Destruction in a Mouse Model of Ross River Virus-Induced Disease," *J. Virol.* 81 (2007), pp. 5132–5143.

7. Mason and Haddow, "An Epidemic of Virus Disease."

8. Rosenthal, "As Earth Warms Up."

9. Rezza, et al., "Infection with Chikungunya Virus."

10. R. K. Pachauri, "Nobel Lecture, December 10, 2007," http://nobelprize.org /nobel_prizes/peace/laureates/2007/ipcc-lecture_en.html (accessed January 8, 2008).

11. K. A. Tsetsarkin, et al., "A Single Mutation in Chikungunya Virus Affects Vector Specificity and Epidemic Potential," *PLoS Pathog.* 3 (2007), e201.

12. Ibid.

13. P. Renault, et al., "Major Epidemic of Chikungunya Virus Infection on Reunion Island, France, 2005–2006," *Am. J. Trop. Med. Hyg.* 77 (2007) pp. 727–731.

14. Tsetsarkin, et al., "A Single Mutation"; and Renault, et al., "Major Epidemic."

15. Rezza, et al., "Infection with Chikungunya Virus."

16. Tsetsarkin, et al., "A Single Mutation."

17. R. N. Charrel, X. de Lamballerie, and D. Raoult, "Chikungunya Outbreaks— The Globalization of Vectorborne Diseases," *New Engl. J. Med.* 356 (2007), pp. 769–771.

18. B. A. Knudsen, "Geographic Spread of *Aedes albopictus* in Europe and the Concern Among Public Health Authorities," *Eur. J. Epidemiol.* 11 (1995), pp. 345–334.

19. C. G. Moore and C. J. Mitchell, "*Aedes albopictus* in the United States: Ten-Year Presence and Public Health Implications." *Emerg. Infect. Dis.* 3 (1997), pp. 329–334.

20. D. Kennedy, "A Tiger Tale," *Science* 297 (2002), p. 1445.

21. Epstein, "Chikungunya Fever Resurgence."

22. R. Sallares, A. Bouwman, and C. Anderung, "The Spread of Malaria to Southern Europe in Antiquity: New Approaches to Old Problems." *Med. Hist.* 48 (2004), pp. 311–328.

23. R. Pinsky, *The Inferno of Dante: A New Verse Translation* (New York: Farrar, Straus & Giroux, 1994), p. xvii.

24. Words in Pictures, Inc., "Ravenna, or How the Glories of Byzantium Ended Up in the Swamps of Italy," http://www.initaly.com/regions/byzant/byzant2.html (accessed January 8, 2008).

25. E. Hutton, *Ravenna, A Study* (New York: E. P. Dutton, 1913), p. 142.

26. R. W. B. Lewis, *Dante* (New York: Penguin/Putnam, 2001).

27. Hutton, *Ravenna.*

28. P. Reiter, "Climate Change and Mosquito-Borne Disease," *Environ. Health Perspect.* 109, Suppl. 1 (2001), pp. 141–161.

29. Rosenthal, "As Earth Warms Up."

30. "Sicurezza alimentare Wwf: sostanze killer nei menu' europei," *Eco News,* http://www.verdi.it/apps/econews.php?id=11002 (accessed January 8, 2008).

31. World Health Organization, "Communicable Disease Surveillance and Response: Chikungunya in Emilia Romagna Region, Italy," http://www.euro.who.int /surveillance /outbreaks/20070904_1 (accessed January 8, 2008).

32. Rosenthal, "As Earth Warms Up."

33. J. P. Chretien and K. J. Linthicum, "Chikungunya in Europe: What's Next?" *Lancet* 370 (2007), pp. 1805–1806.

34. Rosenthal, "As Earth Warms Up."

GIVING THINGS THEIR PROPER NAMES:
CARL LINNAEUS AND W. H. AUDEN

1. C. Linnaeus, (1747) *Wästgöta-Resa.* In: T. Moore, *Unnatural Kinds: Remarks on Linnæus's Classification of Mental Disorders,* http://www.hku.hk/philodep/dept/tm/papers /linnaeus/three2.html (accessed March 2006).

2. W. H. Auden, *A Certain World: A Commonplace Book* (New York: Viking Press, 1970), p. 22.

3. W. H. Auden, *The Age of Anxiety* (New York: Random House, 1947).

4. W. H. Auden, *The Dyer's Hand and Other Essays* (New York: Random House, 1962), p. 81.

5. C. Britton, "King of Flowers," http://www.linnaeus300.com/carl-linnaeus (accessed March 2007).

6. C. Linnaeus, *Philosophia Botanica,* tr. Stephen Freer (Oxford: Oxford University Press, 2005), p. 219; E. O. Wilson, "The Linnaean Enterprise: Past, Present, and Future," *Proc. Am. Phil. Soc.* 149 (2005), pp. 344–349; L. Petrusson, "Carl Linnaeus: Botanical History Swedish Museum of Natural History," http://w2.nrm.se/fbo/hist/linnaeus/linnaeus .html.en (accessed March 2007); G. Broberg, *Homo sapiens:* Linnaeus's Classification of Man. In *Linnaeus: The Man and His Work.* (Berkeley: University of California Press, 1983), pp. 157-194; J. Uglow, *The Lunar Men: Five Friends Whose Curiosity Changed the World* (New York: Farrar, Straus & Giroux, 2002).

7. Linnaeus, *Philosophia Botanica.*

8. Britton, "King of Flowers."

9. Uglow, *The Lunar Men.*

10. R. Porter, *The Enlightenment.* Penguin, New York, p. 273.

11. C. Darwin, *On the Origin of Species.* (New York: Collier, 1909). Chap. 14, p. 9.

12. Broberg, *Homo sapiens.*

13. S. T. Coleridge, *Select Poetry and Prose,* ed. Stephen Potter (London: Nonesuch Press, 1972), pp. 469–470.

14. C. Darwin and A. Wallace, "On the Tendency of Species to form Varieties; and on the Perpetuation of Varieties and Species by Natural Means of Selection," *Proc. Linnean Soc. London,* http://www.linnean.org/index.php?id=53 (accessed March 2007).

15. W. H. Auden, "Voltaire at Ferney." In: *The Collected Poetry of W. H. Auden* (New York: Random House, 1945), p. 6.

16. H. Carpenter, *W. H. Auden: A Biography* (Boston: Houghton Mifflin, 1981), pp. 176–177.

17. R. M. Pettis, "Mann, Erika (1905-1969)," *glbtq: An Encyclopedia of Gay, Lesbian, Bisexual, Transgender, and Queer Culture,* http://www.glbtq.com/arts/mann_e.html (accessed March 2007).

18. W. H. Auden, and L. MacNeice, *Letters from Iceland.* (New York: Random House, 1937).

19. W. H. Auden, "Iceland Revisited." In: *About the House* (London: Faber and Faber, 1966), p. 59.

20. P. Michalzik, *Gustaf Grundgens: Der Schauspieler und die Macht [The Actor and the Powerful]* (Berlin: Koch, 1999).

21. G. Weissmann, "Auden and the Liposome." In: *The Woods Hole Cantata, Essays on Science and Society* (New York: Dodd, Mead, 1985), pp. 79–84.

22. W. H. Auden, "September 1, 1939." In: *The Collected Poetry of W. H. Auden* (New York: Random House, 1945), p. 57.

23. Auden, *The Dyer's Hand.*

24. O. Sacks, "Dear Mr. A . . . ." In: *W. H. Auden: A Tribute.* ed. Stephen Spender (New York: Macmillan, 1975), p. 191.

25. W. H. Auden, "Ode to Terminus," *New York Review of Books,* July 11, 1968.

SPINAL IRRITATION AND FIBROMYALGIA:
LINCOLN'S SURGEON GENERAL AND THE THREE GRACES

1. F. Wolfe, et al., "The American College of Rheumatology 1990 Criteria for the Classification of Fibromyalgia," *Arthritis Rheumat.* 33 (1990) pp. 160–72. [See also http://www.rheumatology.org/publications/classification/fibromyalgia/fibro.asp?aud=mem (accessed December 2007).]

2. W. A. Hammond, *Spinal Irritation* (Detroit: Hammond, 1886).

3. Ibid.

4. Wolfe, et al., "American College of Rheumatology 1990 Criteria."

5. E. Shorter, *From the Mind into the Body: The Cultural Origins of Psychosomatic Symptoms* (New York: Free Press, 1994).

6. R. Staud and M. E. Rodriguez, "Mechanisms of Disease: Pain in Fibromyalgia Syndrome," *Nat. Clin. Pract. Rheumatol.* 2 (2006), pp. 90-98.

7. Ibid.

8. "Clash in Detroit Over How Ill a Kevorkian Patient Really Was," *New York Times,* August 20, 1996, p. A1.

9. Staud and Rodriguez, "Mechanisms of Disease"; and N. Hadler, "'Fibromyalgia' and the Medicalization of Misery," *J. Rheumatol.* 30 (2003) pp. 1668–1670.

10. G. E. Ehrlich, "Fibromyalgia Is Not a Diagnosis," *Arthritis Rheumat.* 48 (2003), p. 276

11. M. B. Yunus, et al., "Electron Microscopic Studies of Muscle Biopsy in Primary Fibromyalgia Syndrome: A Controlled and Blinded Study," *J. Rheumatol.* 1 (1989), pp. 97–101.

12. M. L. Lacroix-Fralish, J. B. Ledoux, and J. S. Mogil, "The Pain Genes Database: An Interactive Web Browser of Pain-Related Transgenic Knockout Studies," *Pain* 131 (2007), pp. e1–e4.

13. L. M. Arnold, et al., "A Double-Blind Multicenter Trial Comparing Duloxetine with Placebo in the Treatment of Fibromyalgia Patients with or without Major Depressive Disorder," *Arthritis Rheumat.* 50 (2004), pp. 2974–2984.

14. L. Färber, et al., "Short-Term Treatment of Primary Fibromyalgia with the 5-HT3-Receptor Antagonist Tropisetron: Results of a Randomized, Double-Blind, Placebo-Controlled Multicenter Trial in 418 Patients," *Intern. J. Clin. Pharmacol. Res.* 21 (2001), pp. 1–13.

15. L. J. Crofford, et al., "Pregabalin for the Treatment of Fibromyalgia Syndrome: Results of a Randomized, Double-Blind, Placebo-Controlled Trial," *Arthritis Rheumat.* 52 (2005), pp. 1264–1273.

16. Ibid.

17. P. Fisher, et al., "Effect of Homeopathic Treatment on Fibrositis (Primary

Fibromyalgia)," *Brit. Med. J.* 299 (1989), pp. 365–366.

18. Ibid.

19. Wolfe, et al., "American College of Rheumatology 1990 Criteria."

20. W. Johnson, *An Essay on the Diseases of Young Women.* (London: Simpkin, Marshall & Co., 1849) pp. 149–151.

21. R. E. Harris, et al., "Comparison of Clinical and Evoked Pain Measures in Fibromyalgia," *J. Pain* 7 (2006), pp. 521–527.

22. J. Giesecke, et al., "Quantitative Sensory Testing in Vulvodynia Patients and Increased Peripheral Pressure Pain Sensitivity," *Obstet. Gynecol.* 104 (2004) pp. 126–133.

23. B. Kellner, *A Gertrude Stein Companion: Content with the Example* (New York: Greenwood Press, 1988), p. 57.

24. M. B. Yunus, S. Arslan, and J. C. Aldag, "Relationship Between Body Mass Index and Fibromyalgia Features," *Scand. J. Rheumatol.* 31 (2002), pp. 27–31.

25. "Clash in Detroit," *New York Times.*

26. SuperStock.com, http://www.superstock.com/ImagePreview/1158-1183 (accessed December 2007).

27. National Galleries of Scotland, http://www.nationalgalleries.org/index.php /collection/online_az/4:322/results/0/38262/ (accessed December 2007).

28. Timeless Myths, http://www.timelessmyths.com/classical/lessergods.html (accessed December 2007).

29. Artchive.com., http://www.artchive.com/artchive/B/botticelli/primavera.jpg .html (accessed December 2007).

30. AnnounceArt.net., http://www.announceart.net/photony/2004/PhotoNY2004 .html (accessed December 2007).

31. Shorter, *From the Mind into the Body,* p. 206.

32. M. C. Gillett, "The Civil War in 1863: Hammond's Last Year." In: *The Army Medical Department 1818–1865.* The Army Medical Department, Center of Military History, Washington. http://history.amedd.army.mil/booksdocs/civil/gillett2/amedd _1818-1865_chpt10.html (accessed December 2007).

33. A. Patten, B. M. Patten, and A. William, "Hammond, the Dynamograph, and Bogus Neurologic Testimony in Old New York," *J. Hist. Neurosci.* 6 (1997), pp. 257–263.

34. Ibid.

35. Hammond, *Spiral Irritation,* p. 260.

36. O. W. Holmes, "The Young and the Old Practitioner." In: *Medical Essays* (Boston: Riverside Press, 1892), p. 378.

TITHONUS AND THE FRUIT FLY: NEW SCIENCE AND OLD MYTHS

1. http://images.google.com/images?ndsp=20&um=1&hl=en&client=firefox-a & channel=s&rls=org.mozilla:en-US:official&hs=zQZ&q=Seymour+Benzer++California +Institute+of+Technology+image&start=0&sa=N.

2. http://www.freud.org.uk/greek.jpg.

3. "Pope: Don't Pursue Immortality Medicine," USA Today.com, http://www.usatoday .com/news/religion/2008-03-09-popeimmortality_N.htm?csp=34.

4. A. P. West, et al., "Crystal Structure of the Ectodomain of Methuselah, a

Drosophila G Protein-coupled Receptor Associated with Extended Lifespan," *Proc. Natl. Acad. Sci. USA* 98 (2001), pp. 3744–3749.

5. E. Lax, *On Being Funny: Woody Allen and Comedy* (New York: Charterhouse, 1975).

6. AP, "Pope: Don't Pursue Immortality Medicine."

7. D. L. Jones, "Aging and the Germ Line: Where Mortality and Immortality Meet," *Stem Cell Rev.* 3 (2007), pp. 192–200; M. Wei, et al., "Life Span Extension by Calorie Restriction Depends on Rim15 and Transcription Factors Downstream of Ras/PKA, Tor, and Sch9," *PLoS Genet.* 4 (2008), p. e13.

8. L. Guarente, G. Ruvkin, and R. Amasino, "Aging, Life Span, and Senescence," *PNAS* 95 (1998), pp. 11034–11036.

9. T. M. Bass, et al., "Effects of Resveratrol on Lifespan in *Drosophila melanogaster* and *Caenorhabditis elegans*," *Mech. Ageing Develop.*. 128 (2007), pp. 546–52.

10. N. Wade, "Scientist at Work: Dr. Leonard Guarente: Searching for Genes to Slow the Hands of Biological Time," *New York Times,* September 26, 2000.

11. http://www.sciencedaily.com/releases/2007/06/070608093844.html.

12. W. W. Ja, et al., "Extension of *Drosophila melanogaster* Life Span with a GPCR Peptide Inhibitor," *Nat. Chem. Biol.* 3 (2007), pp. 415–9.

13. J. R. Lowell, "A Familiar Epistle to a Friend," 1857. In: *The Poetical Works of James Russell Lowell,* Vol. 3 (Boston: Houghton Mifflin, 1896), p. 273.

14. B. Kestenbaum and R. Ferguson, "The Number of Centenarians in the United States on January 1, 1990, 2000, and 2010 Based on Improved Medicare Data," *North Am. Actuarial J.* 10 (2006), pp. 1–6.

15. J. R. Wilmoth, et al., "Increase of Maximum Life-Span in Sweden, 1861–1999," *Science* 289 (2000), pp. 2366–2368.

16. "Chinese Life Expectancy Rises by 41 Years in One Century," http://english .peopledaily.com.cn/english/200010/20/eng20001020_53183.html.

17. W. B. Yeats, "After Long Silence." In: *The Collected Poems of W. B. Yeats* (New York: Macmillan, 1950), p. 260.

18. Ibid., p. 191.

19. Ibid.

20. W. H. Auden, "In Memory of William Butler Yeats." In: *The Collected Poetry of W. H. Auden* (New York: Random House, 1945), p. 48.

21. "Spottiswoode & His Enemies," *That's What I Like* © 2008 Jonathan Spottiswoode, J Spot Music Ascap (700261204321); and http://cdbaby.com/cd/spott5; P. Roth, *The Dying Animal* (Boston, New York: Houghton Mifflin, 2001).

22. S. Freud, "Formulations Regarding the Two Principles in Mental Funtioning." In: *Collected Papers of Sigmund Freud,* Vol. IV (London: Hogarth Press, 1956), p. 18.

23. Ibid.

24. E. Jones, *Sigmund Freud: Life and Work,* Vol. III. (London; The Hogarth Press, 1957) p. 133.

25. A. Lord Tennyson, "Tithonus." In: *Tennyson's Poetical Works* (London: Macmillan, 1891; original in *Cornhill Magazine* 1860), p. 125.

26. Ibid.

27. A. Huxley, *After Many A Summer Dies the Swan* (New York: Harpers, 1939).

28. Ibid. p. 86.

29. pr.caltech.edu/catalog/pdf/catalog_07_08_part1.pdf.

30. West, et al., "Crystal Structure"; T. Fulop, Jr., et al., "Age-Dependent Changes in Transmembrane Signalling: Identification of G Proteins in Human Lymphocytes and Polymorphonuclear Leukocytes," *Cell Signal* 5 (1993), pp. 593–603; and M. R. Philips, et al., "Activation-Dependent Carboxyl Methylation of Neutrophil $G\gamma_2$," *Proc. Natl. Acad. Sci. USA* 92 (1995), pp. 2283–2287.

31. W. Blake, "The Fly." In: *The Poetry and Prose of William Blake,* ed. David V. Erdman (New York: Doubleday, 1965), p. 23.

32. J. Weiner, *Time, Love, Memory: A Great Biologist and His Quest for the Origins of Behavior* (New York: Vintage, 1999).

SWIFTBOATING "AMERICA THE BEAUTIFUL": KATHARINE LEE BATES AND A BOSTON MARRIAGE

1. K. L. Bates, "America the Beautiful," Falmouth Historical Society Archives, http://www.falmouthhistoricalsociety.org/05/klbates.html (accessed September 2007).

2. L. Archer, "Mitt Romney Stumps in Local Tech Circle," http://blog.washingtonpost.com/posttech/2007/04/mitt_romney_stumps_in_local_te_1.html (accessed September 2007).

3. A. Lorentzen, "Iowa Gay Marriage Ruling Stirs 2008 Race," Associated Press, September 1, 2007.

4. Bates, "America the Beautiful."

5. L. E. Richards and M. H. Elliot, *Julia Ward Howe, 1819–1910* (Boston and New York: Houghton Mifflin, 1915).

6. D. Burgess, *Dream and Deed: The Story of Katharine Lee Bates,* (Norman: University of Oklahoma Press, 1952).

7. Bates, "America the Beautiful."

8. R. Bordin, *Alice Freeman Palmer: The Evolution of a New Woman* (Ann Arbor: University of Michigan Press, 1993).

9. Ibid.

10. J. Schwarz, "Yellow Clover: Katharine Lee Bates and Katherine Coman," *Frontiers* 4 (1979), pp. 59–67.

11. Ibid.

12. "The Book of the Fair: The Women's Department 1893," Paul V. Galvin Library, Illinois Institure of Technology, http://columbus.gl.iit.edu (accessed September 2007).

13. "Central Fountain" by Frederick W. MacMonnies, Columbian Exposition. 1893. Paul V. Galvin Library, Illinois Institure of Technology, http://columbus.gl.iit.edu (accessed September 2007).

14. V. Scudder, *On Journey* (New York: E.P. Dutton, 1937).

15. H. Adams, *The Education of Henry Adams* (New York: Penguin Classics, 1995; original 1918), p. 330.

16. G. Cleveland, *Grover Cleveland Speeches; Special Session Message* (August 8, 1893), http://www.millercenter.virginia.edu/scripps/digitalarchive/speeches/spe_1893_0808_cleveland (accessed September 2007).

17. L. W. Knight, *Citizen: Jane Addams and the Struggle for Democracy* (Chicago: University of Chicago Press, 2005), p. 181.

18. S. N. Cleghorn, *Portraits and Protests* (New York: Henry Holt, 1917), p. 56.

19. S. Cosner and J. Scanlon, *American Women Historians* (Westport, CT: Greenwood Press, 1996), p. 44.

20. Ibid.

21. K. L. Bates, *Collected Poetry of Kathlarine Lee Bates: Poems in Memory of Katharine Coman* (Boston & New York: Houghton, Mifflin, 1930), p. 209.

22. Ibid., p. 205.

23. Ibid., p. 189.

NOTHING MAKES SENSE IN MEDICINE
EXCEPT IN THE LIGHT OF BIOLOGY

1. T. Dobzhansky, "Nothing in Biology Makes Sense Except in the Light of Evolution," *Am. Biol. Teacher* 35 (1973), pp. 125–129.

2. C. Bernard, *An Introduction to the Study of Experimental Medicine*, tr. H. C. Green (New York: Henry Schurman, 1875).

3. J. Folkman, "Angiogenesis: An Organizing Principle for Drug Discovery?" *Nature Rev. Drug Discovery* 6 (2007), pp. 273–286.

4. Bernard, *Introduction*.

5. G. Weissmann, "Summary of Symposium." In: *Claude Bernard and the Internal Environment*, ed. E. D. Robin (New York and Basel: Marcel Dekker, 1979) pp. 277–283

6. Bernard, *Introduction*; and P. B. Dunham, et al., "From Beaumont to Poison Ivy: Marine Sponge Cell Aggregation and the Secretory Basis of Inflammation," *Federation Proc.* 44 1985), pp. 2914–2924.

7. Bernard, p. 147.

8. J. Folkman, et al., "Isolation of a Tumor Factor Responsible for Angiogenesis, *J. Exptl. Med.* 133 (1971), pp. 275–288.

9. "Reports" to the editor from High Wire Press, *FASEB J.*

10. Folkman, et al., "Isolation of a Tumor Factor."

11. E. Moschkowitz, "Relation of Lymphocytic Infiltration of Inflammatory Origin to Angiogenesis," *AMA Arch. Pathol.* 49 (1950), pp. 247–266.

12. M. Greenblatt and P. Shubik, "Tumor Angiogenesis: Transfilter Diffusion Studies in the Hamster by the Transparent Chamber Technique," *J. Natl Cancer Inst.* 41 (1968), pp. 111–124.

13. J. Folkman, "Tumor Angiogenesis: Therapeutic Implications," *New Engl. J. Med.* 285 (1971), pp. 1182–1186.

14. J. Folkman, "Anti-Angiogenesis: New Concept for Therapy of Solid Tumors," *Ann. Surg.* 175 (1972), pp. 409–416.

15. Ibid.

16. S. Pincock, "Profile: Judah Folkman: Persistent Pioneer in Cancer Research," *Lancet* 366 (2005) p. 1259.

17. Bernard, *Introduction*, p. 15.

18. Bernard, *Introduction*, p. 78.

19. R. M. Zollinger and E. H. Ellison, "Primary Peptic Ulcerations of the Jejunum Associated with Islet Cell Tumors of the Pancreas," *Ann. Surg.* 142 (1955), p. 709.

20. Pincock, "Profile."

21. Ibid.

22. J. Folkman and L. H. Edmunds, Jr., "Endocrine Pacemaker for Complete Heart Block," *Circulation Res.* 10 (1962), pp. 632–641.

23. J. Folkman and D. M. Long, "The Use of Silicone Rubber as Carrier for Prolonged Drug Therapy," *J. Surg. Res.* 4 (1964), pp. 139–142.

24. S. Segal, *Under the Banyan Tree* (New York: Oxford University Press, 2003), and personal communication.

25. Folkman and Long, "The Use of Silicone Rubber."

26. Segal, *Under the Banyan Tree.*

27. Bernard, *Introduction,* p. 218.

APPLY DIRECTLY TO THE FOREHEAD:
HOLMES, NANA, AND HENNAPECIA

1. O. W. Holmes, "The Young and the Old Practitioner." In: *Medical Essays,* 1892 ed. (Boston: Riverside Press, 1871), p. 378.

2. G. J. Nohynek, et al., "Toxicity and Human Health Risk of Hair Dyes," *Food Chem. Toxicol.* 42 (2004) pp. 517–543; and A. E. Laumann and A. J. Derick, "Tattoos and Body Piercings in the United States: A National Data Set," *J. Am. Acad. Dermatol.* 55 (2006), pp. 413–421.

3. K. Markos, "Scarred Children: Two North Jersey Families Are Suing Over Permanent Marks Their Kids Got from Temporary Tattoos," *The Bergen County Record,* January 5, 2007, p. A01.

4. I. J. Kang and M. H. Lee, "Quantification of Para-phenylenediamine and Heavy Metals in Henna Dye," *Contact Dermatitis* 55 (2006), pp. 26–29.

5. K. A. Abdulla and N. M. Davidson, "A Woman Who Collapsed after Painting Her Soles," *Lancet* 348 (1996), p. 658.

6. J. Hardwicke and S. Azad, "Temporary Henna Tattooing in Siblings—an Unusual Chemical Burn," *Burns* 32 (2006), p. 1064; J. Blair, R. T. Brodell, and S. T. Nedorost, "Dermatitis Associated with Henna Tattoo. 'Safe' Alternative to Permanent Tattoos Carries Risk," *Postgrad. Med.* 116 (2004), pp. 63–65; and M. Onder, "Temporary Holiday 'Tattoos' May Cause Lifelong Allergic Contact Dermatitis When Henna Is Mixed with PPD," *J. Cosmet. Dermatol.* 3 (2003), pp. 126–130.

7. P. Jung, et al., "The Extent of Black Henna Tattoos' Complications Are Not Testricted to PPD-Sensitization," *Contact Dermatitis* 55 (2006), pp. 57–59.

8. Abdulla and Davidson, "A Woman Who Collapsed."

9. J. C. Starr, J. Yunginger, and G. W. Brahser, "Immediate Type I Asthmatic Response to Henna Following Occupational Exposure in Hairdressers," *Ann. Allergy* 48 (1982), pp. 98–99.

10. A. N. Kok, et al., "Henna (*Lawsonia inermis* Linn.) Induced Haemolytic Anaemia in Siblings," *Intern. J. Clin. Pract.* 58 (2204), pp. 530–532.

11. O. Uygur-Bayramicli, R. Dabak and M. Sargin, "Acute Chemical Colitis Resulting from Oral Intake of Henna," *J. Clin. Gastroenterol.* 39 (2005), pp. 920–921.

12. H. Babich and A. Stern, "In Vitro Cytotoxicities of 1,4-Naphthoquinone and Hydro-xylated 1,4-naphthoquinones to Replicating Cells," *J. Appl. Toxicol.* 13 (1993), pp. 353–358.

13. D. C. McMillan, et al., "Role of Oxidant Stress in Lawsone-Induced Hemolytic Anemia, *Toxicol. Sci.* 82 (2004), pp. 647–655.

14. U.S. Food and Drug Administration, "Temporary Tattoos and Henna/Mehndi," 2007, Center for Food Safety and Applied Nutrition, Office of Cosmetics and Colors Fact Sheet, http://www.cfsan.fda.gov/~dms/cos-tatt.html (accessed January 2007).

15. V. H. Price, "Androgenetic Alopecia in Women," *J. Investig. Dermatol. Symp. Proc.* 8 (2003), pp. 24–27.

16. L. Thomas, R. T. McCluskey, and G. Weissmann, "Prevention by Cortisone of the Changes in Cartilage Induced by an Excess of Vitamin A in Rabbits," *Am. J. Pathol.* 42 (1963), pp. 271–283.

17. J. Berth-Jones and P. E. Hutchinson, "Novel Cycle Changes in Scalp Hair Are Caused by Etretinate Therapy," *Brit. J. Dermatol.* 132 (1995), pp. 367–375.

18. P. C. Arck, et al., "Towards a 'Free Radical Theory of Graying': Melanocyte Apoptosis in the Aging Human Hair Follicle Is an Indicator of Oxidative Stress Induced Tissue Damage," *FASEB J.* 20 (2006), pp. 1567–1569.

19. É. Zola, *The Experimental Novel,* tr. Belle M. Sherman (New York: Haskell House, 1964), pp. 2–3.

20. É. Zola, *Nana,* 1880, http://www.gutenberg.org/etext/5250 (accessed January 2007).

21. Ibid.

22. Zola, *The Experimental Novel,* pp. 20–21.

ELIZABETH BLACKWELL BREAKS THE BONDS

1. E. Blackwell, *Pioneer Work in Opening the Medical Profession to Women* (New York: Humanity Books, 2005), pp. 202–230.

2. "Bellevue Hospital Medical College," *New York Times,* March 1, 1867, p. 8.

3. "The Medical Education of Women," *New York Times,* March 1, 1867, p. 8.

4. "Women's Medical College of New York Infirmary for Women and Children," *Leslie's Illustrated Newspaper,* April 16, 1876, p. 1ff.

5. D. Shrier, "A Celebration of Women in U.S. Medicine," *Lancet* 363 (2004). pp. 253–255.

6. Transcript: The Republican Presidential Candidates Debate, *New York Times,* May 3, 2007, http://www.nytimes.com/2007/05/03/us/politics/04transcript.html?ex =133601 (accessed May 2007).

7. K. Demirjian, "Religious Leaders Rip Hate-Crime Measure; House Passes Bill Extending Coverage to Sexual Orientation," *Chicago Tribune,* May 5, 2007, p. 1.

8. Blackwell, *Pioneer Work,* p. 47ff.

9. E. Blackwell, "Ship Fever: An Inaugural Thesis. Submitted for the Degree of M.D., at Geneva Medical College," *Buffalo Medical Journal and Monthly Review* 4 (1849), pp. 523–531

10. H. Blot, "De l'albuminurie chez les femmes enceintes: ses rapports avec l'éclampsie, son influence sur l'hémorrhagie utérine après l'accouchement" (Paris: Rignoux, 1849), p. 358.

11. Blackwell, *Pioneer Work,* p. 178ff.

12. L. A. Altman, "Agency Urges a Change in Antibiotics for Gonorrhea,," *New York Times,* April 13, 2007, p. A10.

13. Blackwell, *Pioneer Work,* p. 194ff.

14. Ibid.

15. Blackwell, *Pioneer Work,* p. 219ff.

16. Blackwell, *Pioneer Work,* p. 239ff.

17. H. B. Elliott, "Woman as Physician." In: *Eminent Women of the Age: Being Narratives of the Lives and Deeds of the Most Prominent Women of the Present Generation,* eds. J. Parton, et al. (Hartford: S. M. Bett & Company, 1869). Quoted in: http://womenshistory.about .com/cs/medicine/a/blackwell_emin.html (accessed May 2007).

CHRONIC LYME DISEASE AND MEDICALLY UNEXPLAINED SYNDROMES

1. R. Leuckart, *Wandtafeln,* http://www.mblwhoilibrary.org/exhibits/leuckart/wall _charts/slide_index.html.

2. A. Bierce, *The Devil's Dictionary,* Alcyone Systems, http://www.alcyone.come /max/lit/devils/.

3. E. Hamilton, "Lyme Disease Guidelines Focus of Antitrust Probe," *Knight Ridder Tribune Business News,* November 17, 2006, p. 1.

4. H. S. Zahran, et al., "Health-Related Quality of Life Surveillance—United States, 1993–2002," *MMWR Surveill. Sum.* 54 (2005), pp. 1–35.

5. Hamilton, "Lime Disease Guidelines."

6. Ibid.

7. G. P. Wormser, et al., "The Clinical Assessment, Treatment, and Prevention of Lyme Disease, Human Granulocytic Anaplasmosis, and Babesiosis: Clinical Practice Guidelines by the Infectious Diseases Society of America," *Clin. Infect. Dis.* 43 (2006), pp. 1089–134.

8. Zahran, et al., "Health-Related Quality of Life Surveillance."

9. Name withheld, letter to the editor, *The Ottawa Citizen,* September 15, 2006, p. B4.

10. E. Shorter, *From Paralysis to Fatigue: A History of Psychosomatic Illness in the Modern Era* (New York: Free Press, 1993).

11. Ibid.

12. S. Wessely, "Medically Unexplained Symptoms: Exacerbating Factors in the Doctor-patient Encounter," *J. Roy. Soc. Med.* 96 (2003), pp. 223–227.

13. P. Henningsen, T. Zimmermann, and H. Sattel, "Medically Unexplained Physical Symptoms, Anxiety, and Depression: A Meta-analytic Review," *Psychosomat. Med.* 65 (2003), pp. 528–533.

14. L. Rangel, et al., "The Course of Severe Chronic Fatigue Syndrome in Childhood," *Proc. Roy. Soc. Med.* 93 (2000), pp. 129–134.

15. Henningsen, et al., "Medically Unexplained Physical Symptoms."

16. Zahran, et al., "Health-related Quality of Life Surveillance."

17. S. Nettleton, "'I just want permission to be ill': Towards a Sociology of Medically Unexplained Symptoms," *Soc. Sci. Med.* 62 (2006), pp. 1167–1178.

18. Reuters, "Briton Cures Fatigue by Drilling Hole in Own Head," February 22, 2000.

19. Ibid.

EUGENICS AND THE IMMIGRANT: ROSALYN YALOW AND RITA LEVI-MONTALCINI

1. R. Yalow, *The Prix Nobel. The Nobel Prizes 1977*, ed. W. Odelberg, Nobel Foundation, Stockholm. http://nobelprize.org/medicine/laureates/1977/yalow-autobio .html (accessed June 2006).

2. R. Levi-Montalcini, *In Praise of Imperfection* (New York: Basic Books, 1988), pp. 117–122.

3. Thomson Scientific, "Science Watch Study Shows United States Loses Dominant Share of World Science," press release, http://scientific.thomson.com/press/2005/8282889 (accessed June 2006).

4. E. Dutt, "Scientists Denied U.S. Visa; Prof. Mehta, Prominent Scientist, Applied for a Visa and That Is Being Issued," *News India-Times,* March 3, 2006, p. 17.

5. C. Barrett, "Why America Needs to Open Its Doors Wide to Foreign Talent," *Financial Times,* January 31, 2006, p. 19.

6. J. A. Mervis, "Glass Ceiling for Asian Scientists?" *Science* 310 (2005), pp. 606–607.

7. K. Pearson and M. Moul, "The Problem of Alien Immigration into Great Britain, Illustrated by an Examination of Russian and Polish Jewish Children," *Ann. Eugenics* (now *Ann. Human Genetics*) 1 (1925), pp. 1–128.

SCIENCE IN THE MIDDLE EAST:
ROBERT KOCH AND THE CHOLERA WAR

1. S. Erlanger, "Haifa Copes as Paralysis of Wartime Takes Hold: Stressful Conditions Prevail at Shelters," *International Herald Tribune,* July 28, 2006, p. 5; and University of Haifa, http://www.haifa.ac.il/index.html.en (accessed August 2006).

2. American University of Beirut, http://www.aub.edu.lb/challenge/notices/depts/oira-20060731.html (accessed August 2006).

3. T. D. Brock, *Robert Koch* (Berlin/New York: SpringerVerlag, 1988), p. 60.

4. Ibid., p. 154.

5. P. de Kruif, *Microbe Hunters* (New York: Harcourt Brace, 1926), p. 124.

6. Ibid.

7. Ibid.

8. Ibid.

9. P. de Kruif, *Arrowsmith* (New York: Harcourt Brace, 1925), p. 243.

10. J. Loeb, *The Organism as a Whole* (New York: G. Putnam's Sons, 1916), p. iv.

HOW TO WIN A NOBEL PRIZE:
THINKING INSIDE AND OUTSIDE THE BOX

1. E. Dickinson, *Complete Poems,* ed. T. H. Johnson (Boston: Little, Brown, 1960), p. 1331.

2. Craig C. Mello, interview with Adam Smith, Nobel Foundation, http://nobelprize.org/nobel_prizes/medicine/laureates/2006/mello-telephone.html (accessed November 2006).

3. G. Weissmann, "Lewis Thomas 1913–1993," *Biogr. Mem. Natl. Acad. Sci. USA* 85 (2004) pp. 314–347.

4. R. D. Kornberg, "Mediator and the Mechanism of Transcriptional Activation," *Trends Biochem. Sci.* 30 (2005), pp. 235–239.

5. Arthur Kronberg, Nobel Lecture, Nobel Foundation, http://nobelprize.org/nobel_prizes/chemistry/laureates/2006/kornberg-telephone.html.

6. Ibid.

7. R. Hooke, *Preface to Micrographia,* (1665. repr., New York: Dover, 1962), p. xi.

8. A. Fire, et al., "Potent and Specific Genetic Interference by Double-Stranded RNA in *Caenorhabditis elegans,"* *Nature* 391 (1998), pp. 806–881.

9. Andrew Z. Fire, interview with Adam Smith, Nobel Foundation, http://-/nobelprize.org/nobel_prizes/medicine/laureates/2006/fire-telephone.html (accessed November 2006).

10. G. L. Sen and H. M. Blau, "A Brief History of RNAi: The Silence of the Genes," *FASEB J.* 20 (2006), pp. 1293–1299.

11. C. Napoli, C. Lemieux and R. Jorgensen, "Introduction of a Chimeric Chalcone Synthase Gene into Petunia Results in Reversible Co-Suppression of Homologous Genes in Trans," *Plant Cell* 2 (1990), pp. 279–289.

12. Ibid.

13. Ibid.

14. The Nobel Prize in Physiology or Medicine 1934. Nobelprize.org, http://www.nobel.se/medicine/laureates/1934/ (accessed November 2006).

15. H. K. Beecher and M. D. Altschule, *Medicine at Harvard: The First 300 Years* (Dartmouth, NH: New England Universities Press, 1977), p. 304.

16. G. S. Whipple and F. S. Robscheit-Robbins, "Blood Regeneration in Severe Anemia. Favorable Influence of Liver, Heart and Skeletal Muscle in Diet," *Am. J. Physiol.* 78 (1925), pp. 408–418.

17. G. R. Minot and W. P. Murphy, "Observations on Patients with Pernicious Anemia Partaking of a Special Diet: A Clinical Aspect," *Trans. AssOc. Am. Phys.* 41 (1926), pp. 72–75.

18. W. P. Castle, "The Conquest of Pernicious Anemia." In: *Blood Pure and Eloquent* ed. M. Wintrobe (New York: McGraw-Hill, 1980), p. 297.

HOMER SMITH AND THE LUNGFISH: THE LAST GASP OF INTELLIGENT DESIGN

1. H. Smith, "The Evolution of the Kidney." In: *Homer William Smith: His Scientific and Literary Achievements,* eds. Herbert Chasis and William Goldring (New York: New York University Press, 1965), p. 81.

2. J. T. Bridgham, S. M. Carroll, and J. W. Thornton, "Evolution of Hormone-receptor Complexity by Molecular Exploitation," *Science* 312 (2006), pp. 97–101. See also http://www.physorg.com/news64046019.html.

3. C. O'Brien, as quoted in *New York Times,* April 7, 2007, sect. 4, p. 2.

4. United States District Court for the Middle District of Pennsylvania, Transcript of Proceedings: Tammy Kitzmiller, et al. v. Dover Area School District, et al., Day 11, (2005). Access Research Network, http://www.arn.org/docs/dovertrial/behe3_2005_1018_day11_am.pdf (accessed April 2007).

5. L. Colombo, et al., "Aldosterone and the Conquest of Land," *J. Endocrinol. Invest.* 29 (2006) pp. 373–381.

6. Ibid.; and Bridgham, et al., "Evolution of Hormone-Receptor Complexity."

7. S. Brownback, "Reject Embryo Bill," *USA Today,* April 10, 2007, p. A10.

8. M. K. Richardson and G. Keuck, "Haeckel's ABC of Evolution and Development," *Biol. Rev. Cambridge Phil. Soc.* 77 (2002), pp. 495–528.

9. Ibid.

10. Ibid.

11. B. Harrub, "Haeckel's Hoax–CONTINUED!" Apologetics Press.org., 2001, http://www.apologeticspress.org/articles/2049 (accessed April 2007).

12. 4 Hurting Christians, Feb. 2006 (2006) update, http://www.4hurtingchristians .com/evidence_of_why_charles_darwin_is_wrong_about_evolution.html (accessed April 2007).

13. ValiantForTruth, "Darwin was Wrong," April 6, 2007, http://valiantfortruth .townhall.com/g/eaecb933–41a1–491a-9626-86ab78a6f890 (accessed April 2007).

14. J. Sarfati and M. Matthews, "Irreducible Complexity." In: *Refuting Evolution 2,* AnswersinGenesis.org, http://www.answersingenesis.org/home/area/re2/chapter10.asp (accessed April 2007).

15. C. Darwin, *The Descent of Man* (1871; repr. Norwalk, CT: Eaton Press, 1974), p. 140.

16. Smith, "The Evolution of the Kidney."

17. Ibid.

18. I. Dinesen, "The Dreamers." In: *Seven Gothic Tales* (New York: Smith & Haas, 1943), p. 178.

19. H. Smith, *From Fish to Philosopher: The Story of Our Internal Environment* (Boston: Little Brown, 1953), p. 4.

20. Ibid., p. 77.

21. J. M. Joss, "Lungfish Evolution and Development," *Gen. Comp. Endocrinol.* 148 (2006), pp. 285–289.

22. T. R. Gregory, "Animal Genome Size Database" (2007). http://www.genomesize .com (accessed April 2007).

23. A. Einstein, "Introduction." In: *Man and His Gods* (New York: Little Brown, 1952), p. 1.

DDT IS BACK: LET US SPRAY!

1. Editorial: "Let Us Spray," *The Sunday Telegraph,* September 17, 2006, p. 20.

2. A. Mandavilli, "DDT Returns," *Nat. Med.* 12 (2006) pp. 870–871.

3. C. W. Dugger, "W.H.O. Supports Wider Use of DDT vs. Malaria," *New York Times,* September 16, 2006, p. A7.

4. A. Attaran and R. Maharaj, "Ethical Debate: Doctoring Malaria, Badly: The Global Campaign to Ban DDT," *Brit. Med. J.* 321 (2000), pp. 1403–1405.

5. Ibid.

6. E. Trepman, "Rescue of the Remnants: The British Emergency Medical Relief Operation in Belsen Camp 1945." *J. Roy. Army Med. Corps* 147 (2001), pp. 281–293

7. Nobelprize.org. The Nobel Prize in Physiology or Medicine 1948. http://www.nobel .se/medicine/laureates/1948 (accessed September 2006

8. M. Satchell and D. L. Boroughs, "Rocks and Hard Places. DDT: Dangerous Scourge or Last Resort?" *US News & World Report* 129 (2000), pp. 64–65.

9. R. Carson, *Silent Spring* (Cambridge. MA: Houghton Mifflin (Riverside Press), 1962). pp. 7–8.

10. W. J. Rogan and A. Chen, "Health Risks and Benefits of Bis(4-chlorophenyl)-1,1,1-trichloroethane (DDT)," *Lancet* 366 (2005,) pp. 763–773.

11. J. L. Bast, P. J. Hill and R. C. Rue, *Eco-Sanity: A Common-Sense Guide to Environmentalism* (Lanham, MD: Madison Books, 1994), p. 100.

12. M. L. Mabaso, B. Sharp, and C. Lengeler, "Historical Review of Malarial Control in Southern African with Emphasis on the Use of Indoor Residual House-Spraying," *Trop. Med. Intern. Health* 9 (2004), pp. 846–856.

13. C. Guinovart, et al., "Malaria: Burden of Disease," *Curr. Mol. Med.* 6 (2006), pp. 37–40.

14. Centers for Disease Control and Prevention, "Fact Sheet: Dengue and Dengue Hemorrhagic Fever" (2003), http://www.cdc.gov/ncidod/dvbid/dengue/facts.html (accessed September 2006).

15. World Health Organization, "Global Health Situation and Projections Estimates: Infectious and Parasitic Diseases: Trypanosomiasis" (1992), http://www.ciesin.columbia .edu/docs/001-010/001-010c.html (accessed September 2006).

16. D. Raoult, et al., "Jail Fever (Epidemic Typhus) Outbreak in Burundi," *Emerg. Infect. Dis.* 3 (1997), pp. 357–360.

17. K. Mokrani, et al., "Reemerging Threat of Epidemic Typhus in Algeria," *J. Clin. Microbiol.* 8 (2004), pp. 3898–3900.

18. R. Taylor and A. Rieger, "Medicine as a Social Science: Rudolf Virchow on the Typhus Epidemic of Upper Silesia," *Intern. J. Health Serv.* 15 (1985), pp. 547–559.

19. Ibid.

20. M. Müller, *Anne Frank: The Biography,* tr. R. Kimber, (New York: Henry Holt, 1998)

21. Exil-Club, "Anne Frank" (2001), http://www.exilclub.de/html/30_projekte/31 _projekte_00/biografien/annefrank/15ende.html (accessed September 2006).

ACADEMIC BOYCOTTS AND THE ROYAL SOCIETY

1. "Boycotting Universities: Slamming Israel, Giving Palestinians a Free Pass," *The Economist,* June 14, 2007, p. 68.

2. E. B. Davies, "Royal Society Opposes Academic Boycott of Israeli Scientists" (2006), http://www.mth.kcl.ac.uk/staff/eb_davies/boycott.doc (accessed August 2007).

3. C. Webster, "New Light on the Invisible College: The Social Relations of English Science in the Mid-Seventeenth Century," *Trans. Roy. Hist. Soc.* 5th Ser., 24 (1974), pp. 19–42.

4. R. H. Syfret, "Some Early Reactions to the Royal Society," *Notes Records Roy. Soc. London* 7 (1950), pp. 7207–7258.

5. S. Usborne, "Class Action: Academics Against Israel," *The Independent,* June 8, 2007, p. 1.

6. University College Union, "Congress Resolution on Israel/Palestine: Arrangements for Implementation" (July 4, 2007), http://www.ucu.org.uk/circ/html/ucu31.html (accessed August 2007); and G. Alderman, "Boycott, Shmoycott," *Jerusalem Post,* July 4, 2007, p. 13.

7. Webster, "New Light."

8. The Royal Society, "Royal Society Opposes Academic Boycott of Israeli Scientists" (May 30, 2006), http://www.royalsoc.ac.uk/news.asp?id=4772 (accessed August 2007).

9. The Elie Wiesel Foundation for Humanity, "Open Letter" (June 11, 2007), http://www.eliewieselfoundation.org/ (accessed August 2007).

10. The American Physiological Society, "APS Responds to Proposed Boycott of Israeli Academics" (June 27, 2007), http://www.the aps.org/pa/resources/archives/comments/boycottobjection.htm (accessed August 2007).

11. Associated Press, "More Than 10,000 Scholars from Across the Political Spectrum and Around the World Add Their Names to Statement Expressing Outrage at British University and Colleges Union Vote to Promote Academic Boycott of Israel," PR Newswire, http://biz.yahoo.com/prnews/070730 /clm039.html?.v=89 (accessed August 2007).

12. Lee C. Bollinger, "Statement by President Lee C. Bollinger on British University and College Union Boycott," Columbia University (June 12, 2007), http://www.columbia .edu/cu/news/07 /06/boycott.html (accessed August 2007).

13. UCU Congress Resolution.

14. Alderman, "Boycott, Schmoycott."

15. R. B. Standler, "History of Academic Freedom," http://www.rbs2.com/afree .htm-anchor111111 (accessed August 2007).

16. F. Engels, "Dialectics of Nature" (1886), http://www.marxists.org/archive/marx /works/1883/don/appendix2.htm (accessed August 2007).

17. E. Haeckel, *Freie Wissenschaft und Frei Lehre* (Stuttgart: E. Schweizerbart, 1878).

18. F. R. Johnson, "Gresham College: Precursor of the Royal Society," *J. Hist. Ideas* 1 (1940), pp. 413–438.

19. H. Hartley and C. Hinshelwood, "Gresham College and the Royal Society," *Notes Records Roy. Soc. London,* 16 (1961), pp. 125–135.

20. Johnson, "Gresham College."

21. H. Trevor-Roper, *The Crisis of the Seventeenth Century: Religion, the Reformation and Social Change* (Indianapolis: Liberty Fund, 2001).

22. G. S. McIntyre, "Brouncker, William, Second Viscount Brouncker of Lyons (1620–1684)." In: *Oxford Dictionary of National Biography* (London: Oxford University Press, 2004).

23. R. Boyle, *The Sceptical Chymist* (London: J. Caldwell, 1661), p. 243.

24. R. Hooke, *Micrographia* (1665; repr. New York: Dover, 1962), p. 86.

25. Ibid., 142, Observation XVIII.

26. G. B. Evans, "The Source of Shadwell's Character of Sir Formal Trifle in 'The Virtuoso,'" *Mod. Lang. Rev.* 35 (1940), pp. 211–214.

27. Dobell, C., *Antony Van Leeuwenhoek and his "Little Animals"* (London: Russel & Russel, 1958).

TEACH EVOLUTION, LEARN SCIENCE: JOHN WILLIAM DRAPER AND THE "BONE BILL"

1. F. S. Collins, *The Language of God* (New York: Simon & Shuster, 2006), p. 12.

2. J. D. Watson, *DNA: The Secret of Life* (New York: Alfred A. Knopf, 2002), p. xii.

3. T. Dobzhansky, "Nothing in Biology Makes Sense Except in the Light of Evolution," *Am. Biol. Teacher* 35 (1973), pp. 125–129.

4. J. D. Miller, E. C. Scott, and S. Okamoto, "Science Communication: Public Acceptance of Evolution," *Science* 313 (2006), pp. 765–766.

5. Ibid.

6. Editorial, "Revival in Iran," *Nature* 442 (2006), pp. 719–720.

7. G. Liles, "God's Work in the Lab," *MD Magazine* 32 (1992), pp. 49–53.

8. Collins, *The Language of God.*

9. C. Venter, "Sequencing the Human Genome: Lecture at the Marine Biological Laboratory, Woods Hole, Massachusetts, August 17, 2001, http://www.mblwhoilibrary .org/services/lecture_series/venter/ (accessed August 2006).

10. Watson, *DNA: The Secret of Life.*

11. S. Brenner, "Hunting the Metaphor," *Science* 291 (2001), pp. 1265–1266.

12. Miller, et al., "Science Communication."

13. Editorial. *Harper's New Monthly Magazine* 8 (1853–1854), p.690

14. D. Fleming, *John William Draper* (Philadelphia: University of Pennsylvania Press, 1950).

15. J. W. Draper, *The History of the Conflict Between Religion and Science* (New York: Appleton, 1874), p. vi.

16. J. R. Wilmoth, "The Future of Human Longevity: A Demographer's Perspective," *Science* 280 (1898), pp. 395–397.

17. Fleming, *John William Draper.*

18. J. W. Draper, "On the Production of Light by Heat," *Am. J. Sci. Arts* 4 (1847), pp. 388–402.

19. K. S. Thomson, "Huxley, Wilberforce and the Oxford Museum," *Am. Sci.* 88 (2002), pp. 210–215.

20. O. Chadwick, *The Secularization of the European Mind in the 19th Century* (Cambridge: Cambridge University Press, 1975), p. 162.

21. K. Marx, "Zur Kritik der Hegelschen Rechtsphilosophie. Einleitung." In: *Marx and Engels: Basic Writings on Politics and Philosophy*, ed. Lewis S. Feuer (New York: Doubleday Anchor, 1844), p. 262.

DIDEROT AND THE YETI CRAB: THE ENCYCLOPEDIAS OF LIFE

1. D. Diderot, *Encyclopédie ou Dictionnaire raisonné des sciences, des arts et des metiers* (1751), http://www.lib.uchicago.edu/efts/ARTFL/projects/encyc/ (accessed June 2007). See also Horace, *Ars Poetica*, v 240, http://meta.montclair.edu/spectator/text/june1711 /ars_poetica.html. (accessed June 2007).

2. The Encyclopedia of Life, Mission Statement (2007), http://www.eol.org/home .html (accessed June 2007). See also E. O. Wilson, *Trends Ecol. Evol.* 18 (2003), pp. 77–80.

3. Dr. Seuss. *Horton Hears a Who* (New York: Random House, 1954), p. 6.

4. S. Brownback, "What I Think About Evolution," *New York Times,* May 31, 2007, p. A19.

5. R. A. Gibbs, et al., "Rhesus Macaque Genome Sequencing and Analysis Consortium, Evolutionary and Biomedical Insights from the Rhesus Macaque Genome." *Science* 316 (2007), pp. 222–234.

6. T. Ackerman, "Scientist Unlocks Secrets in Own DNA: First Map of Human Genome Paves the Way for Personalized Care," *Houston Chronicle,* June 1, 2007, p. 1.

7. Diderot, *Encyclopédie.*

8. E. O. Wilson, *Trends Ecol. Evol.* 18 (2003), pp. 77–80.

9. The Encyclopedia of Life, Mission Statement.

10. Wilson, *Trends.*

11. Ibid.

12. The Encyclopedia of Life, Mission Statement.

13. Ibid.

14. Universal Biological Indexer and Organizer, MBLWHOI Library, http://www .ubio.org/ (accessed June 2007).

15. Brownback, "What I Think About Evolution."

16. The Encyclopedia of Life, Mission Statement.

17. Diderot, *Encyclopédie.*

18. A. M. Wilson, *Diderot* (New York: Oxford University Press, 1972), pp. 232ff.

19. Diderot, *Encyclopédie.*

20. The Encyclopedia of Life, Mission Statement.

21. R. Darnton, *The Great Cat Massacre, And Other Episodes in French Cultural History* (New York: Random House, 1985), p. 298.

22. Ibid.

23. D. Diderot, "*Le Rêve d'Alembert.*" In: *Le Neveu De Rameau (*Paris: Cartonnage éditeur Le Club Francais du Livre, 1947), p. 317.

24. D. Diderot, *Suite d'Apologie de M l'Abée de Prades* (1752) [cited in Wilson, *Diderot,* p. 170].

25. G. Weissmann, "Titanic and Leviathan," *Oceanus* 31 (1988), pp. 69–77.

26. A. Nevala, "Lurking Benignly on the Seafloor, the 'Yeti' Crab Is Discovered," (2006), *Oceanus,* http://www.whoi.edu/oceanus/viewArticle.do?id=12327 (accessed June 2007).

27. E. MacPherson, W. Jones, and M. Segonzac, "A New Squat Lobster Family of Galatheoidea (Crustacea, Decapoda, Anomura) from the Hydrothermal Vents of the Pacific-Antarctic Ridge," *Zoosystema* 27 (2005), pp. 709–723.

28. Universal Biological Indexer and Organizer, MBLWHOI Library, Yeti Crab (2006), http://www.ubio.org/browser/details.php?namebankID=5994978 (accessed June 2007).

DENGUE FEVER IN RIO: MACUMBA VERSUS VOLTAIRE

1. "Brazil: Dengue Toll Hits 92 in Rio State," www.washingtonpost.com/wp-dyn /content/article/2008/04/24/AR2008042400141.html.

2. M. Margolis, "Dengue Plagues Rio," *Newsweek,* April 14, 2008.

3. Voltaire, *Candide* (Paris: Nouveau Classiques Larousse, 1970). Also, http://www .literature.org/authors/voltaire/candide/.

4. F. Giubu, "Minister of Health Says Brazil Has the War Against Dengue," http://www1 .folha.uol.com.br/folha/cotidiano/ult95u397543.shtml.

5. M. Astor, "Faulty Response to Fever Outbreak Plaguing Brazil: Dengue Fatalities Climb, Especially in Young Children, *Houston Chronicle,* April 13, 2008, p. 19.

6. Ibid.

7. R. S. Rodriguez-Tan and M. R. Weir, "Dengue: A Review," *Texas Med.* 94, no.10 (1998), pp. 53–59.

8. Ibid.

9. S. Thein, et al., "Changes in Levels of Anti-Dengue Virus IgG Subclasses in Patients with Disease of Varying Severity," *J. Med. Virol.* 40 (1993), pp. 102–106.

10. Rodriguez-Tan and Weir, "Dengue: A Review"; and P. Avirutnan, et al., "Dengue Virus Infection of Human Endothelial Cells Leads to Chemokine Production, Complement Activation, and Apoptosis," *J. Immunol.* 161 (1998), pp. 6338–6346.

11. AP, "Brazil War Against Dengue Fever Slowed," http://www.usatoday.com /news/world/2008-04-10-dengue-brazil_N.html.

12. "In Times of Dengue," www.esculhambacao.com.br, *Quinta-feira*, April 17, 2008.

13. G. S. E. Simson, *Black Religions in the New World* (New York: Columbia University Press, 1978).

14. "In Times of Dengue."

15. S. Booker, "Santeria: The Beliefs and Rituals of a Growing Religion in America," *Afri. Stud.Rev.* 48 (2005), pp. 199–201.

16. G. MacEoin, "U.S. Troops to El Salvador," *National Catholic Reporter,* October 13, 2000.

17. J. A. Rawlings, et al., *Am. J. Trop. Med. Hy*g., 59, no. 1 (July 1998), pp. 95–99.

18. *Virus Weekly,* "Dengue Fever: Mosquito-Borne Virus an Emerging Public Health Problem in U.S.," November 7, 2000, http://www.newsrx.com/newsletters/Virus-Weekly /2000-11-07/2000110733312RW.html.

19. D. G. Mcneil, Jr., "Hovering Where Rich and Poor Meet, the Mosquito," *New York Times*, September 3, 2000, p. D4.

20. AP, "Brazil War Against Dengue Fever Slowed."

21. S. Grudgings (Tue.). "Campbell Adds Voice, Not Blood, to Rio Dengue Fight," Reuters, April 15, 2008, www.reuters.com/article/latestCrisis/idUSN15476572

22. N. Lange, "A macumba da dengue," *Observatório da Imprensa*, www.observato-riodaimprensa.com.br/artigos /fd060320021.html.

23. "Mutirão de médicos homeopatas atende crianças com dengue no Rio." *Folha de S. Paulo.* April 14, 2008.

24. E. Nauman, "Dengue Fever Alert. Homeopathy: 21st Century Medicine," Access: NewAge, http://www.accessnewage.com/articles/health/DENGUE.HTM.

25. Lange, "A macumba da dengue" ; and "Mutirão de médicos," 26. J. Jacobs, et al., "The Use of Homeopathic Combination Remedy for Dengue Fever Symptoms: A Pilot RCT in Honduras," *Homeopathy* 96 (2007), pp. 22–26.

27. Leonard Bernstein, *Candide: A Comic Operetta Based on Voltaire's Satire*, lyrics by Richard Wilbur, Stephen Sondheim Lillian Hellman, et al., 1956, http://www .sondheimguide.com/Candide/56libretto1-2o.html#One:2.

28. See note 3, chap. 6.

29. Margolis, "Dengue Plagues Rio"; and Giubu, "Minister of Health Says."

30. Voltaire, *Lettres philosophies,* ed. Gustave Lanson, 2 vols. (Paris: Hachette, 1924), i. p. 74.

31. A. Ardao, "Assimilation and Transformation of Positivism in Latin America," *J. Hist. Ideas*, 24 (1963), pp. 515–522.

32. G. Lenzer, *Auguste Comte and Positivism: The Essential Writings* (Chicago: University of Chicago Press, 1975), pp. 317–18.

33. See note 3, chap. 18.

34. F. B. Dos Santos, et al., "Recombinant Polypeptide Antigen-Based Immuno-globulin G Enzyme-Linked Immunosorbent Assay for Serodiagnosis of Dengue," *Clin Vaccine Immunol.* 14 (2007), pp. 641–643.

Permissions

Mortal and Immortal DNA: Craig Venter and the Lure of "Lamia"
Page 13 (left): From Albertus Seba, *Cabinet of Natural Curiosities* (1734–1765)
Page 13 (right): "Lamia," from Edward Topsell's *History of Fourfooted Beasts* (1607).
BOTH IMAGES COURTESY OF THE MBLWHOI LIBRARY (WOODS HOLE, MA, USA)

Homeopathy: Holmes, Hogwarts, and the Prince of Wales
Page 19: Oliver Wendell Holmes, M.D. 1880 lithograph, photographer unknown.
COURTESY PRIVATE COLLECTION

Citizen Pinel and the Madman at Bellevue
Page 25: Tony Robert-Fleury (1838–1911). 1876 painting: *Citizen Pinel orders removal of the chains of the mad at the Salpêtrière.* RÉUNION DES MUSÉES NATIONAUX/ART RESOURCE, NY
Page 29: Philippe Pinel (1745–1826). *Projections of the normal skull.* From *A Treatise on Insanity,* 1801. COURTESY PRIVATE COLLECTION

The Experimental Pathology of Stress: Hans Selye to Paris Hilton
Page 32 (left): Hans Selye (1907–1982). PHOTO COURTESY OF THE AMERICAN INSTITUTE OF STRESS
Page 32 (right): Paris Hilton PHOTO BY FREDERICK M. BROWN/GETTY IMAGES
Page 36 : Calciphylaxis: Cutaneous molt induced by calciphylaxis in the rat. (Selye, H., Gentile, G., Prioreschi, P. Calciphylaxis: Cutaneous molt induced by calciphylaxis in the rat. *Science,* 1961, 134:1876-1877). REPRINTED WITH PERMISSION FROM AAAS

Gore's Fever and Dante's *Inferno*: Chikungunya Reaches Ravenna
Page 39: Canto XIX of Dante Alighieri's *Inferno;* print, after Sandro Botticelli.
SNARK/ART RESOURCE, NY
Page 41: The tiger mosquito: *Aedes albopictus.* COURTESY OF THE NORTH CAROLINA DEPARTMENT OF ENVIRONMENTAL AND NATURAL RESOURCES, DIVISION OF ENVIRONMENTAL HEALTH

Giving Things Their Proper Names: Carl Linnaeus and W. H. Auden
Page 45 (left): Carl Linnaeus (1707–1778). Engraving by Ehrensvärd, ca. 1740.
Page 45 (right): W. H. Auden, with Erika Mann, 1935.
PHOTO BY A.D. BANGHAM WITH PERMISSION
Page 48 : Photocopy of a marriage register. A. D. BANGHAM, WITH PERMISSION

Spinal Irritation and Fibromyalgia: Lincoln's Surgeon General and the Three Graces
Page 53 (left): The 18 "Pressure Points" of Fibromyalgia. (Wolfe, F., Smythe, H.A., Yunus, M.B., et al. The American College of Rheumatology 1990 Criteria for the Classification of Fibromyalgia. *Arthritis Rheum.* 1990 Feb; 33, 160–172). REPRINTED WITH PERMISSION OF WILEY-LISS, INC., A SUBSIDIARY OF JOHN WILEY & SONS, INC.
Page 53 (right): William A. Hammond (1828–1900) Surgeon General, U.S. Army, 1862–4. COURTESY OF THE MBLWHOI LIBRARY (WOODS HOLE, MA, USA)

Tithonus and the Fruit Fly: New Science and Old Myths
Page 60 (left): Seymour Benzer (1921–2007), with wooden model of *Drosophila* (ca. 1974). COURTESY OF THE ARCHIVES, CALIFORNIA INSTITUTE OF TECHNOLOGY
Page 60 (right): Vase: Eo pursuing Tithonus. COURTESY OF THE FREUD MUSEUM, LONDON

Swift-boating "America the Beautiful": Katharine Lee Bates and a Boston Marriage
Page 67 (left): "The Republic" (1893), by Daniel Chester French.
COURTESY OF THE PAUL V. GALVIN LIBRARY, ILLINOIS INSTITUTE OF TECHNOLOGY
Page 67 (right): Katharine Lee Bates (1859–1929), poet of "America the Beautiful."
COURTESY OF WELLESLEY COLLEGE ARCHIVES; PHOTO BY PARTRIDGE.
Page 70: "Central Fountain of the Columbian Exposition," by Frederick W. Macmonnies
(Chicago 1893). COURTESY OF THE PAUL V. GALVIN LIBRARY, ILLINOIS INSTITUTE OF TECHNOLOGY

Nothing Makes Sense in Medicine Except in the Light of Biology
Page 75 (left): *The Lesson of Claude Bernard* (1884) by Léon Augustin L'Hermitte.
BRIDGEMAN-GIRAUDON/ART RESOURCE, NY
Page 75 (right): Judah Folkman (1933–2008). COURTESY OF CHILDREN'S HOSPITAL, BOSTON;
PHOTO BY BACHRACH

Apply Directly to the Forehead: Holmes, Zola, and Hennapecia
Page 81 (left): Dr. Oliver Wendell Holmes (1809–1894).
Page 81 (right): *Nana* (Emile Zola). Stamp issued by La Poste (2003).
BOTH IMAGES COURTESY PRIVATE COLLECTION

Elizabeth Blackwell Breaks the Bonds
Page 86 (left): Albert Lewis Sayre at Bellevue Hospital Medical College. "Spinal Traction"
(1876). COURTESY OF THE NYU SCHOOL OF MEDICINE EHRMAN MEDICAL LIBRARY ARCHIVES
Page 86 (right): Women's Medical College of New York Infirmary *(Leslie's Illustrated
Newspaper,* April 16, 1876). COURTESY OF THE NATIONAL LIBRARY OF MEDICINE
Page 92: Elizabeth Blackwell, M.D., ca. 1890.
COURTESY OF THE NATIONAL LIBRARY OF MEDICINE

Chronic Lyme Disease and Medically Unexplained Syndromes
Page 94: Rudolph Leuckart: "*Arthropoda*" Wallchart I.
COURTESY OF THE MBLWHOI LIBRARY (WOODS HOLE, MA, USA)

Eugenics and the Immigrant: Rosalyn Yalow and Rita Levi-Montalcini
Page 100 (left): Rosalyn Yalow (Prix Nobel, 1977). AP IMAGES
Page 100 (right): Rita Levi-Montalcini (Prix Nobel, 1986).
IMAGE COURTESY OF BECKER MEDICAL LIBRARY, WASHINGTON UNIVERSITY SCHOOL
OF MEDICINE
Page 102: Karl Pearson (*left*), FRS (1857–1936), and Francis Galtonx (*right*), FRS at age 81
(1822–1911)
IMAGE COURTESY OF GALTON.ORG, HTTP://WWW.GALTON.ORG/

Science in the Middle East: Robert Koch and the Cholera War
Page 105: On the fiftieth anniversary of the death of Robert Koch, 27 May 1960.
COURTESY BUNDESPOST, WIKIPEDIA COMMONS (PUBLIC DOMAIN)

How to Win a Nobel Prize: Thinking Inside and Outside the Box
Page 110 (left): Roger Kornberg.
Page 110 (right): Arthur Kornberg BOTH PHOTOS BY A. D. BANGHAM, WITH PERMISSION
Page 113: Petunia flowers exhibiting sense co-suppression (RNAi) patterns of chalcone syn-
thase silencing. *Plant Cell Online* by Richard Jorgensen. Copyright 1990 by American
Society of Plant Biologists. REPRODUCED WITH PERMISSION OF AMERICAN SOCIETY OF PLANT
BIOLOGISTS IN THE FORMAT TRADEBOOK VIA COPYRIGHT CLEARANCE CENTER

Homer Smith and the Lungfish: The Last Gasp of Intelligent Design
Page 117 (left): Chordates Come Ashore. From *Kunstformen der Natur*, Ernst Haeckel
(ca. 1902). COURTESY OF THE MBLWHOI LIBRARY (WOODS HOLE, MA, USA)
Page 117 (right): Homer Smith (1895–1962). Physiologist and author of *Kamongo: The
Lungfish and the Padre* (1932) COURTESY PRIVATE COLLECTION
Page 120: The African Lungfish (*Protopterus aethiopicus aethiopicus*).
COURTESY OF THE MBLWHOI LIBRARY (WOODS HOLE, MA, USA)

DDT Is Back: Let Us Spray!
Page 123: Rudolph Leuckart: "*Arthropoda*" (1880) Wallchart II.
  COURTESY OF THE MBLWHOI LIBRARY (WOODS HOLE, MA, USA)

Academic Boycotts and the Royal Society
Page 128 (left): Robert Hooke "The Flea" in Micrographia (1661).
Page 128 (right): Robert Boyle's *Sceptical Chymist* (1661 ed.).
  BOTH IMAGES COURTESY PRIVATE COLLECTION

Teach Evolution, Learn Science: John William Draper and the "Bone Bill"
Page 135: FASEB motto 2007.
  COURTESY *The FASEB Journal,* http/www.fasebj.org/cgi/content/full/20/13/2183/F2
Page 138: John William Draper, M.D. (1811–1882) COURTESY PRIVATE COLLECTION

Diderot and the Yeti Crab: The Encyclopedias of Life
Page 141 (left): Denis Diderot (1713–1784), of the *Encyclopédie* (1751). Portrait by L.-M. van Loo, 1767 (Louvre). RÉUNION DES MUSÉES NATIONAUX/ART RESOURCE, NY
Page 141 (right): "Yeti Crab," *Kiwa Hirsuta,* of the *Encyclopedia of Life* (2007).
  IMAGE ©IFREMER/A. FIFI; COURTESY OF THE MBLWHOI LIBRARY (WOODS HOLE, MA, USA)
Page 145: Frontispiece of the 1772 edition of the *Encyclopédie;* drawn by Charles-Nicolas Cochin, engraved by Bonaventure-Louis Prévost. COURTESY PRIVATE COLLECTION

Dengue Fever in Rio: Macumba versus Voltaire
Page 149 (left): Rudolph Leuckart: "*Arthropoda*" Wallchart III.
  COURTESY OF THE MBLWHOI LIBRARY (WOODS HOLE, MA, USA)
Page 149 (right): Voltaire (1689–1778). COURTESY RARE BOOKS DIVISION, NEW YORK PUBLIC LIBRARY, ASTOR, LENOX AND TILDEN FOUNDATIONS.

Index

Page references in *italics* refer to illustrations.

Remsen, David, 143
"Republic, The" (French), *67*
Rêve d'Alembert (Diderot), 145
Rich, Jessica, 33
Richardson, Albert, 58
Riefenstahl, Leni, 51
RNA, 13–14, 112, 113
Robert-Fleury, Tony, *25, 29*
Roberts, Richard, 14, 61
Robscheit-Robbins, Frieda, 115
Rockefeller University, 79
Roman expérimental, Le (Zola), 84
Roosevelt, Franklin Delano, 101
Roosevelt, Theodore, 110
Rose, Charlie, 15
Rosellini, Roberto, 44
Rostand, Jean, 9
Roth, Philip, 63
Rousseau, Jean-Jacques, 144
Roux, Emile, 107–108, 109
Royal Society of London, 113, 128, 129,
    131–132, 133–134
Rubens, Peter Paul, 56

S
Sacks, Oliver, 51
"Sailing to Byzantium" (song), 63
"Sailing to Byzantium" (Yeats), 62–63
Sanitary Commission, 92
Santeria, 151
Sayre, Albert Lewis, *86,* 87
Sceptical Chymist, The (Boyle), *128,* 132
schizophrenia, 26–28
Schoenheimer, Rudolf, 100
Science, 36, 117, 136, 142
Science Daily, 61
scrofula, 20
Seba, Albertus, 46, 48
Segal, Sheldon, 79–80
Segonzac, Michel, 147
self renewal, asymmetric, 15
Selling the Code (Selye), 38
Selye, Hans, *32,* 34, 35–38
Sen, George L., 113
settlement-house movement, 72
Seuss, Dr., 141
Shadwell, Thomas, 133
Sharp, Philip A., 14
Sherard, James, 46
Sherley, Jim, 15

Shorter, Edward, 57, 96, 98
Shubik, Philippe, 77
Silent Spring (Carson), 123, 124, 125
sleeping sickness, 124
Smith, Adam, 9, 110–111
Smith, Homer, *117,* 120–121, 122
Smithies, Oliver, 13, 14, 16
Smithsonian Institution, 142
Smoot, George F., 110
Snow, C. P., 11
Snow, John, 108
South, Robert, 131, 132
spectrum analysis, 139
Spender, Stephen, 51
spinal irritation, 53–59
Spinal Irritation (Hammond), 53, 57
"Spinal Traction," *86*
Spitzer, Elliot, 26
Spottiswoode, Jonathan, 63
St. Bartholomew's Hospital, 90
Stahl, Frank, 14
Stanton, Edwin M., 57
Stanton, Elizabeth Cady, 71
Starr, Ellen Gates, 72
Stein, Gertrude, 56
stem cells, 13, 14, 15–16, 118, 136, 138
steroids, 36, 80
Straus, Isador, 107
stress, 32–38, 60, 61, 118
"Stresses and Strains of Homeostasis"
    (Cannon), 37
strikes, 72, 73
Sunday Telegraph, The, 123
Sun (ligand), 61, 66
Sussman, Simon, 100
Systema Natura (Linnaeus), 46, 47

T
Tajbakhsh, Shahragim, 13, 16
Tamminga, Carol, 27
Tanay, Emanuel, 27
Tarantino, Anthony, 147
Tarloff, David, 25–26
tatoos, 82, 83
taxonomy, binomial, 47
Taylor, D. J., 32
telegraphy, 139
Temple of Nature, The (Darwin), 48
Temporão, Jose Gomes, 149–150
Temps, Le, 107